Routes of Compromise

THE MEXICAN EXPERIENCE

William H. Beezley, series editor

ROUTES of COMPROMISE

COMPROMISE

Building Roads and Shaping the Nation
in Mexico, 1917–1952   MICHAEL K. BESS

University of Nebraska Press · Lincoln & London

Portions of chapters 2, 3, and 4 originally appeared in "Revolutionary Paths: Road Building, National Identity, and Foreign Power in Mexico, 1917–1938," in *Mexican Studies/ Estudios Mexicanos* 32, no. 1 (Winter 2016): 56–82. Portions of chapters 5 and 6 and the maps of Nuevo León and Veracruz originally appeared in "Routes of Conflict: Building Roads and Shaping the Nation in World War II and Postwar Mexico, 1941– 1952," in *Journal of Transport History* 35, no. 1 (June 2014): 78–96.

Library of Congress Cataloging-in-Publication Data
Names: Bess, Michael Kirkland, author.
Title: Routes of compromise: building roads and shaping the nation in Mexico, 1917–1952 / Michael K. Bess.
Description: Lincoln: University of Nebraska Press, [2017] | Series: The Mexican experience | Includes bibliographical references and index.
Identifiers: LCCN 2017017418 (print)
LCCN 2017022856 (ebook)
ISBN 9780803299344 (cloth: alk. paper)
ISBN 9781496202468 (pbk.: alk. paper)
ISBN 9781496204011 (epub)
ISBN 9781496204028 (mobi)
ISBN 9781496204035 (pdf)
Subjects: LCSH: Roads—Mexico—Design and construction—History—20th century. | Highway engineering—Mexico—History—20th century. | Transportation, Automotive—Government policy—Mexico—History—20th century.
Classification: LCC TE28 (ebook)
LCC TE28 .B47 2017 (print)
DDC 338.4/76257097209041—dc23
LC record available at https://lccn.loc.gov/2017017418

Set in Whitman by Rachel Gould.

For Mercedes and Jeanette

CONTENTS

FIGURES

ACKNOWLEDGMENTS

Writing a book is both an individual and a collective effort. With that idea in mind, I would like to thank Samuel Brunk for his guidance and attention to my work as it developed into the book you hold in your hands. I am also deeply indebted to the examples of Ernesto Chávez and Charles Ambler as writers, scholars, and teachers. Likewise Laura Shelton, Matthew O´Brien, Michael Fitzgerald, Robert Batchelor, Craig Roell, Stacey Sowards, Robert Doyle, Michelle Haberland, Adam Aaronson, Sandra McGee-Deutsch, Paul Edison, Jeff Shepherd, Debra Sabia, Donald Rakestraw, Jon Bryant, Nancy Shumaker, and Jon Steinberg have, in different ways, inspired and encouraged me in this endeavor.

Over the years many friends have provided valuable advice as well as good humor, patience, and opportunities to occasionally step away from the keyboard; they include especially Danielle Smith, Lina Murillo, Kristopher Klein, Dennis Aguirre, Susie Aquilina, Jaime Ruíz, and Heather Sinclair. New friends and colleagues at the Centro de Investigación y Docencia Económicas, including Catherine Vézina, Luis Barrón, Andrew Paxman, Pablo Mijangos, Catherine Andrews, Raul Pacheco-Vega, Gabriel Purón, David Juárez, and Miguel Rodrigues, have been a great source of support, reviewing chapters and calculations, as I developed this manuscript. Michael Ducey, at the Universidad Veracruzana, helped me to gain access to local sources in Veracruz that I had worried would not be available. In addition, my research assistants, Ana Fernanda Fraga, Berenice Hernández, and Daniela Ivonne Herrara de la Cruz, aided me in covering even more ground in Mexico City's historical collections at the Archivo General de la Nación, the Archivo Historico de la Suprema Corte de la Justicia de la Nación, the Fideicomiso Archivos Plutarco Elías Calles y Fernando Torreblanca, and the Biblioteca Lerdo de Tejada. I would like to thank all of the dedicated archivists I have had the opportunity to work with at the Archivo General de la Nación, the Archivo General del Estado de Veracruz, the Archivo General del Gobierno de Nuevo León, the Archivo Histórico de PEMEX, the Hemeroteca Nacional de la Universidad Nacional Autónoma de México, and the Hemeroteca de la Universidad Autónoma de Nuevo León.

I am grateful for the feedback and help provided by William Beezley, Gordon Pirie, Benjamin Smith, Andrew Grant Wood, Susan Gauss, Barry Carr, Jürgen Buchanaeu, Dhan Zunino Singh, Massimo Moraglio, Kyle Shelton, Anna Alexander, and the anonymous reviewers, as well as Bridget Barry, my editor; Joeth Zucco, my project manager; and Emily Wendell at the University of Nebraska Press. I am also thankful to Judith Hoover, my copyeditor, for the careful attention she paid to my work. Comments from audiences at conferences hosted by Transport, Traffic and Mobility (T2M), the Rocky Mountain Council of Latin American Studies, the Western Historical Association, and the American Historical Association have been indispensable in thinking about and revising this project. My research received generous funding through multiple scholarships, in particular the Diana Natalicio Doctoral Dissertation Fellowship and the George A. Krutilek Memorial Graduate Fellowship at the University of Texas at El Paso, as well as a North American Mobility grant from Georgia Southern University.

Finally, I would like to express my deep gratitude for the love and reassurance of my family: to my partner and best friend, Mayeli (who also helped with calculations and developing graphs); my parents, Nora and Melvin; my cousins, Steven, JR, Hazel, Debbie, Jon, Miguel, and their families. Special thanks to Ryan Otero, Lynn Bruder, Eva Hernández, Hayden Vaverek, Mike Kukula, Carmen Hernández, Ryan Larson, Skip Holden, Andrew Fredericks, Isaac Pérez Bolado, Crystal and Jessica Peralta, Robert Harland, la Casa de los Amigos, and Mike Edwards.

ABBREVIATIONS

AMC	Asociación Mexicana de Caminos (Mexican Road Association)
CCE	Comisión de Caminos del Estado (Commission for State Roads)
CD	Cámara de Diputados (Chamber of Deputies)
CIPO	Compañía Internacional de Petróleo y Oleoductos (International Company for Petroleum and Pipelines)
CLC	Comisión Local de Caminos (Local Road Commission)
CNCV	Comisión Nacional de Caminos Vecinales (National Commission for Local Roads)
COP	Comunicaciones y Obras Públicas (Communications and Public Works)
CROM	Confederación Regional Obrera Mexicana (Regional Confederation of Mexican Labor)
CSOCV	Confederación Sindicalista de Obreros y Campesinos del Estado de Veracruz (Union Confederation of Laborers and Farmers of Veracruz State)
CTM	Confederación de Trabajadores de México (Confederation of Mexican Workers)
DCOP	Departamento de Comunicaciones y Obras Públicas (Department of Communications and Public Works)

DFA	Departamento de Fomento y Agricultura (Department of Development and Agriculture)
DNC	Dirección Nacional de Caminos (National Road Directorate)
DSI	Dirección de Servicios de Investigación y Análisis (Directorate of Research and Analysis Services)
FL	Fondo Legal (Legal Collection)
FP	Fondos Presidenciales (Presidential Collections)
HPC	Huasteca Petroleum Company
INEGI	Instituto Nacional de Estadística y Geografía (National Institute of Stastistics and Geography)
IP	Informes Presidenciales (Presidential Reports)
JCCV	Junta Central de Caminos de Veracruz (Central Road Board of Veracruz)
JLC	Juntas Locales de Caminos (Local Road Boards)
JLCNL	Junta Local de Caminos de Nuevo León (Local Road Board of Nuevo León)
PRI	Partido Revolucionario Institucional (Institutional Revolutionary Party)
SCJN	Suprema Corte de la Justicia de la Nación (Supreme Court of Justice of the Nation)
SCOP	Secretaría de Comunicaciones y Obras Públicas (Secretariat of Communications and Public Works)
SEGOB	Secretaría de Gobernación (Secretariat of the Interior)
SEOCC	Sindicato de Empleados y Obreros Constructores de Caminos (Union of Employees and Construction Workers for Roads)
SEOCCENL	Sindicato de Empleados y Obreros Constructores de Caminos de Nuevo León (Union of Employees and Road Workers of Nuevo León)
SEP	Secretaria de Educación Pública (Ministry of Public Education)

Routes of Compromise

Introduction

Revolutionary Roads

On 1 September 1918, President Venustiano Carranza sat on the dais in the ornate Donceles Legislative Palace in the heart of Mexico City to deliver his annual address to the Chamber of Deputies. "Highways deserve special attention," he told the gathering, "it is absolutely necessary that the country has a complete road network." He urged the assembly to rebuild the nation's transportation infrastructure, which years of armed civil conflict following the 1910 Revolution had significantly damaged. Twelve months later he expounded on this need: "The social reconstruction of the nation is made manifest . . . in the repair of old roads, in improvements to Mexico City's streets, in the re-opening of closed thoroughfares, and in the building of new ones."[1]

Carranza did not live long enough to see his ambition for road building fully realized. By 1919 construction plans had stalled due to political strife and strained finances, and in May 1920 an assassin took the president's life. More than a year passed before the federal government, now under his rival Álvaro Obregón, attempted a nationwide program to build new motor highways. The country faced an enormous challenge. Mexico in 1921 counted fewer than 28,000 kilometers of roads for a landmass that covered 2 million square kilometers. Most of these routes were rudimentary dirt and gravel paths that could not accommodate motor traffic. In comparison, at this time the United States had already built more than

100,000 kilometers of "auto-trails" with more than 8 million registered motor vehicles in circulation.[2]

Mexicans responded to the paucity of motorways by forming pro-road committees to advocate for construction efforts. They urged the government to invest millions of pesos in new engineering projects. Every sector of society—including workers, farmers, teachers, landowners, petty officials, business people, state bureaucrats, local and regional *caciques* (political bosses), governors, judges, and federal administrators—participated in the debate about how the country's road network should be formed.

In doing so the nation grappled with significant technical obstacles. The engineers and laborers who built the first motor highways and *caminos vecinales* (local roads) encountered harsh conditions in remote locations with difficult terrain and bad weather that threatened their work. Gravel and asphalt-concrete routes could be built at much steeper grades than railroads, but they remained susceptible to heavy springtime rains that limited motor vehicle access to mountainous communities.[3]

National construction efforts had to consider the legal and practical hurdles of working around existing built environments (homes, farms, and ranches, for instance). New motorways altered the sense of space in a given place, affecting everyday life in different ways. On the one hand, a new road meant the possibility of greater economic and social mobility thanks to easier access to bus routes and reduced travel times to markets to sell local goods. On the other hand, it could lead to destruction of arable land and the loss of property and potential future earnings. As a result, people contested road building fiercely. Wealthy landowners and local caciques filed lawsuits and used subordinates to block projects they disliked. Many communities relied on civic organizations, labor unions, and agrarian networks, as well as the courts, to implore public officials to provide redress.

The presidents of local road-building committees became influential actors in shaping political support for construction efforts. In a sense, drawing on Alan Knight's discussion of *caciquismo* in the twentieth century for perspective, I argue that they acted as "road-building caciques" who facilitated (or stymied) the official bureaucracy.[4] These individuals mediated the demands of workers, farmers, and other citizens with the objectives laid out by the state governor and engineering agencies for new transportation infrastructure. Local opposition to road-building proposals could occur strategically in order to extract more favorable terms from the government. In this way ordinary people participated in state formation by contesting the scope and impact of large-scale infrastructure projects. The pragmatic compromises that the authorities brokered with road-building caciques and other interest groups to complete motorways underscored the reality of these endeavors as *public* works.

Historicizing Road Building as State Formation

In this book I explore the social, economic, political, and legal implications of road building as a critical aspect of state formation. Construction efforts not only manifested state power in large-scale infrastructure projects but also helped to mobilize public and private resources to once isolated areas. James C. Scott and Eugen Weber have written on the role that road-building efforts have played historically in shaping a sense of national identity as people traveled more easily from place to place. New roads joined regional markets into national economies as trade links extended beyond state borders, forging new avenues for international commerce and contact.[5]

Despite state claims to greater technical efficiency, road building was often politically messy, fraught with myriad social and legal complications that spanned national and local interests. In studying the case of Mexico, my work helps to pull apart the notion of a monolithic national state, highlighting how aspects of it were appropriated by special interests through legal means or otherwise. I identify how underlying processes of "disorder" in everyday construction efforts (acrimonious legal disputes, engineering delays, strikes and protests, corruption, abuses of power, bad weather) affected modernist projects that later became monuments to "orderliness" and government effectiveness in glossy tourism brochures and state propaganda. The routes that construction brigades etched across the landscape, moving southward from the Mexican-U.S. borderlands and outward from Mexico City to the Pacific and Gulf coasts, profoundly affected society.

The following chapters are based on case studies drawn from comparative analysis of Nuevo León and Veracruz, two states that served as critical trade corridors for Mexico to U.S. and overseas markets. By the start of the twentieth century, Nuevo León's capital, Monterrey, was a major manufacturing center that drove industrialization in the northeastern part of the country and forged commercial linkages with Texas. The northern oil fields in the state of Veracruz were a crucial export zone, while the port cities of Veracruz and Tuxpan were major targets of infrastructure spending to build new highways to connect them to Mexico City.

These states shared characteristics that made them economically important to the nation and exemplified competing models for road building. From the 1920s onward, commercial and political elites in Nuevo León favored top-down planning and private-sector participation to develop and build motorways. In contrast, in Veracruz agrarian and labor groups organized grassroots movements in conjunction with radical state politicians to empower rural communities to join in construction efforts. The evolution of these competing models over three decades is an important thread throughout this book.

I analyze the Mexican state through the prism of its engineering bureaucracies, studying how these agencies functioned as social organizations that not only addressed technical issues but also engaged with local communities over their concerns for motorway infrastructure. A regional approach underscores the powerful influence of state officials and local groups vis-à-vis federal authorities in the decision-making process for new roads. No single political figure or central agency entirely dominated this process. Instead provincial road-building efforts highlighted the *contingent* nature of political power in Mexico.[6]

The study of state formation has long been an important part of the literature on twentieth-century Mexico. It emerged partly as a reaction to the debate between official and revisionist histories of the Mexican Revolution, which tended to frame it as the modern country's formative "event." Beginning in the 1980s and early 1990s scholars pressed for a more comprehensive understanding of the revolutionary epoch. Gil Joseph, Daniel Nugent, Mary Kay Vaughan,

Alan Knight, and others have argued that this history should be studied as a process wherein changes to the Mexican state took shape over many years after President Porfirio Díaz's long rule, a period known as the Porfiriato (1876–1910), ended.[7]

Work by James Scott, Ranajit Guha, and Antonio Gramsci influenced this appreciation of the Revolution as a historical space where the "low politics" of campesino and labor movements engaged (or clashed) with the "high politics" of the revolutionary caudillos and other political elites who built roads and the state. Scholars of Mexican state formation have critiqued earlier historiographic views that failed to appreciate the political and social motivations of nonelite rural and urban Mexicans. Regional case studies have sought to capture a clearer idea of what motivated the ordinary citizens who rose up against President Díaz, joined revolutionary armies, and helped to forge the new state that came after.[8]

Likewise these historians have cautioned against writing about "the state" in monolithic terms. Quoting the sociologist Philip Abrams, Joseph and Nugent argue, "We should abandon the state as a material object of study whether concrete or abstract while continuing to take the *idea* of the state extremely seriously. . . . The state is, then, in every sense of the term a triumph of concealment. It conceals the real history and relations of subjection behind an a-historical mask of legitimating illusion. . . . In sum: the state is not the reality which stands behind the mask of political practice. It is itself the mask."[9] They urge scholars to critically examine the state as a process of *formation*, dissecting its constituent components and internal contradictions—in this sense, peeling back the mask reveals the social history of the institutions that survived (or emerged from) the Revolution. As in the case of road building in Mexico, it encourages us to delve into the organizational structure of the agencies, and their personnel, which gave shape to the state.[10] By doing so we chip away at the monolithic notion of a hegemonic entity capable of easily overpowering its opponents in favor of a more nuanced appreciation of how governance is riven with discord and compromise, both within and outside the halls of officialdom.

The creation of a nationwide road network unfolded as a series of overlapping political and bureaucratic activities addressing spatial and socioeconomic issues with significant local interests at play. Federal planners in Mexico City looked to new motor roads as a means to develop the country's economy, connecting key population centers and promoting greater commercial activity. Whereas national leaders guided the broad vision for this endeavor, it was often at the state and local levels where public officials, agency planners, and community leaders debated how best to realize the specific details of new roadwork.[11]

The scholarly literature on Mexican road building has emphasized the role these efforts played in state formation. Wendy Waters finds significant local participation, whereby rural communities engaged with state officials and local caciques to build motorways. Her work highlights the grassroots perspective of this history as citizens engaged with the state in its myriad forms to carve out communal benefits in the form of improved access to regional transport mobility. The national government guided these efforts through broad policy initiatives, which were then handed off to and interpreted by state and municipal authorities for implementation. While some local communities opposed road construction, others came to identify new roads as tangible symbols of the nation emerging out of years of revolutionary strife.[12]

Aspects of nationalism and national identity are closely tied to road building. J. Brian Freeman captures these characteristics in his work on the political uses of motorways as symbols of modernization in Mexico. He studies the automobile races that ran along the Pan-American Highway during the 1950s, which national promoters hoped would showcase Mexican engineering achievements for international audiences. The reporting of these events actually ran counter to this expectation; instead of emphasizing positive exposure of the country, foreign media focused on the spectacular motor accidents that occurred. The government worried that this coverage tarnished Mexico's image globally and pulled its support for future races. Freeman has also written on how individual citizens responded to the social changes and economic opportunities that came with new automotive technologies. He finds that Mexico City transportation unions formed in the 1910s, creating a new kind of industrial worker: bus and taxi drivers.[13]

Roads and access to motor travel changed how people perceived their country, making driving tours a popular activity with those who had the means to own automobiles. Dina Berger and Andrew Grant Wood, as well as Freeman, are some of the scholars in the growing field of tourism studies, who have examined this history in Mexico.[14] They identify the relationship among key political themes, including Pan-Americanism, President Franklin Delano Roosevelt's Good Neighbor Policy, and the Mexican government's modernization program. Tourism improved U.S.-Mexico relations and became a convenient source of income that complemented road-building efforts as driving clubs from California and Texas ventured southward. By the 1930s improved domestic political stability was a boon for the tourism industry. New highways brought Americans to Carnival in Veracruz and exposed them to Mexico's pre-Columbian heritage as the newly formed Instituto Nacional de Antropología y Historia excavated important sites like Teotihuacán. Tourist money also flowed into casino and hotel schemes and the vice industry along the border and in coastal cities.[15]

My work is the first to fully take on a study of the institutional and legal implications of road building at the national, state, and local levels.[16] I scrutinize the activities of the federal *juntas locales de caminos* (local road boards) that directed regional construction efforts. These agencies worked alongside state and local entities, including pro-road-building committees that local citizens set up to have a voice in the negotiations and implementation of new routes. When disputes emerged, higher-ranking officials intervened, and often the courts became the final arbiters of who controlled access to road funds and labor.

The size and complexity of the bureaucracy charged with construction efforts made it a target for unscrupulous actors.[17] Corruption took many forms: locally citizen complaints included allegations of misappropriated resources, forced land sales with inadequate remuneration, unfair taxes to pay for faraway roads, stolen road worker wages, and abuse of power by committee bosses; at state and national levels, there arose cases of collusion over federal contracts, conflicts of interest between public officials and the private sector, quid pro quo arrangements, bribes, and trading of inside information, among other issues.[18] Abuses reached the upper echelons of the bureaucracy, where prominent figures such as Juan Andreu Almazán, Maximino Ávila Camacho, and President Miguel Alemán became enmeshed in corruption accusations.

Amid these challenges the courts played an instrumental role in ensuring open access to the nation's motorways. They often sided with federal and state governments on expropriation cases, ordering property owners to cede tracts of land for new roads and not to impede construction efforts. Reforms to federal road laws empowered local communities and state legislatures to sue for unrestricted access to high-quality thoroughfares private firms built to facilitate industrial operations. Local petitioners based their claims on article 11 of the Constitution of 1917, which said that "every man has a right to enter the Republic, exit it, travel through its territory, and change his residence without the need of a security card, passport, or any similar device." The courts vigorously defended this right of free access. In the 1930s the Suprema Corte de la Justicia de la Nación (Supreme Court of Justice of the Nation, scjn), Mexico's highest court, frequently ruled against writs of *amparo* (*writ of protection*) that individuals and companies brought forward to restrict public access on privately built roads. By denying these claims the courts ensured that motorways were seen as a public good available for anyone to use.[19]

The Social Landscape of Road Building

Driving cars and riding in motorbuses changed how many Mexican citizens experienced their country. New roads collapsed travel times, connecting rural communities with nearby towns and cities, and in the process created economic opportunities. Low fares made the cost of motor travel affordable for many people. In cities professionals and workers crowded onto busy urban buses that charged five cents for rides, while rural cooperatives repurposed flatbed trucks to inexpensively carry goods and passengers. Moreover bus drivers formed workers' unions deeply influenced by the rhetoric of domestic and international labor movements.[20]

In the 1920s and early 1930s federal and state governments did not clearly delineate by type when they touted the building of new motorways. Dirt, macadam, and asphalt-concrete routes received the "modern" label in official reports on the progress of construction efforts. This situation created confusion and considerable frustration among the public as dirt roads became impassable for motor vehicles after inclement weather. In response many rural towns and regional cities took the lead in calling for improvements to surface quality. In letters to the government, citizens described poor driving conditions and the lack of regular bus service as acutely detrimental to their material well-being. The government responded to these concerns by issuing public bonds to finance more roads and increasing the construction budgets of federal and state agencies. By the mid-1940s a notable policy shift had occurred in the kinds of roads being built, with a majority of asphalt-concrete projects now under way.

In this sense road building and surface quality became metrics that national society used to gauge national "progress." Motorways carried a dual narrative: on the one hand, they embodied the promise of technological modernization as footpaths and horse-drawn carts gave way to asphalt thoroughfares and automobiles; on the other hand, the effects of entropy due to use, weather, and the passage of time could transform these "modern" arteries into symbols of "backwardness," decried by journalists and citizens if not properly maintained. Political leaders were sensitive to the condition of the road system in their jurisdiction, viewing it as symbolic of their own political legacies.[21]

Road building in early to-mid-twentieth-century Mexico occurred at a time of considerable interest worldwide in this kind of engineering. In the United States during the 1910s newspapers and civic associations boasted about the Lincoln Highway, the first all-weather transcontinental motorway, which extended 5,454 kilometers from New York to California. For many Americans it epitomized the country's technical prowess; the New York Times characterized U.S. road building and the automotive industry as models for other nations to follow.[22]

Across Latin America road building paralleled state formation efforts as citizens formed associations that raised funds for new motorways and lobbied government authorities to participate. In Chile the tourism industry and middle-class citizens drove interest in highways as the nation looked to the United States for design influences and new equipment through technology transfers. Argentina by the early 1940s had constructed more than 30,000 kilometers of new roads, a 1,400 percent increase that had occurred over two decades. Road building in Brazil strengthened the central state vis-à-vis regional elites, while many Brazilians, as occurred in other places around the world, began to equate national progress with the quality of the country's roads.[23]

As many national road-building projects got under way, plans for transnational infrastructures also coalesced. The Lincoln Highway inspired the Meridian Highway, which would extend from Canada to Mexico and later became known as the Pan-American Highway. The promise of economic development through tourism and trade buoyed motorway ambitions across the hemisphere. By the 1930s and 1940s pro-road groups in California and western Mexico touted the Pacific Highway as an important binational public works project even as they disagreed on the route it should take from Los Angeles to Mexico City.[24]

The Mexican press regularly construed road building as a mechanism to enhance the nation's economic prosperity and improve its social character. A 1925 essay in the Mexico City newspaper *Excélsior* argued that new roads helped to "promote and spread general and individual well-being" and served as "moralizing, educative, and civilizing influences for the good of society."[25] In another editorial it reiterated this point, calling for "every man, woman, and child of Mexico to fight for good highways."[26] The newspaper attributed notions of physical vigor to the construction of motorways. For proponents new roads and motor travel served as tangible examples of how the country's technological progress subdued the natural environment and reordered physical space into quantifiable units legible on an engineer's map.

Technology, State Planning, and the Nation

The history of building transportation infrastructure as emblematic of a larger narrative about modernity and progress has deep roots in Mexico. During the Porfiriato national officials saw railroad construction as part of a "civilizing mission" that attracted foreign investment and developed export markets for commodities, like copper, that further fueled domestic industrialization. While foreigners certainly benefited from this arrangement, regional studies of railroad carriage rates reveal that as much as 70 percent of goods in circulation went to domestic markets. This economic activity ignited fierce public debate across Mexico, embracing railways as a vehicle to modernize the nation, while also questioning whether the government was adequately addressing locomotive accidents and labor abuses.[27]

Coordinating policy efforts to cope with the broader economic impact of railroads was a chief concern for domestic officials and business leaders. In Monterrey in the 1880s and 1890s the deployment of new railways initially had a damaging impact on the city's industrial productivity. Railroads facilitated the importation of lower-cost goods from the United States, while also reducing the cost for skilled artisans to relocate to other parts of the republic. It was not until new federal protectionist measures went into effect after 1899—which shielded local industries in Nuevo León from foreign competition—that the city was finally able to effectively utilize railroads to drive regional economic expansion.[28]

Similar preoccupations pertained to policies that governed the building of motor roads. During the 1920s national leaders set rules limiting the role of foreign companies in this work and stipulated that Mexican engineers and laborers receive hiring preference. In 1925 President Plutarco Elías Calles touted these guidelines as a means to ensure that road building remained a nationalistic endeavor, putting Mexico's needs ahead of foreign interests.[29] In many ways this policy worked well; highway construction employed thousands of citizens as laborers and also trained a new generation of domestic engineers.

Yet, just as railroad historians have emphasized, construction budgets for new motorways were not boundless. They required prioritization of public works that helped some communities and ignored others. Big cities like Mexico City, Monterrey, and the port of Veracruz benefited greatly from official attention that placed them at the center of highway designs, while many smaller communities waited years for new transport infrastructure to arrive. In letters that petitioned for aid in support of local roads rural towns and hamlets in remote locations complained of having "been forgotten" by the government.[30] As with railroad building, new motor routes successfully integrated key regional markets into Mexico's national economy but did not necessarily reduce problems of social, economic, and spatial inequality.

Nevertheless inequality is only one part of this story. Studying the long-term economic impact of road building on everyday life in Mexico reveals a correspondingly complex social image. National and state elites did not benefit alone to the detriment of all other groups. Rather, fierce competition often emerged at local and state levels to influence construction efforts. Leaders in towns and smaller cities recognized that the particular route a road or highway took through their *municipios* (counties) could have a profound impact on their individual and communal fortunes.

Far from being passive or disinterested subjects, nonelite citizens actively participated in road planning, often as enthusiastic consumers of motor transportation. This point finds parallels with railroad travel, where growing numbers of poor and working-class individuals had frequently used third-class passenger service.[31] While wealthy Mexicans and foreigners profited from motorways, they were not alone in taking advantage of this infrastructure. Regular bus service complemented new roads, facilitating rural trade with faraway markets and providing greater access to health care, education, and other public services. New roads were not divorced from everyday life, but rather became deeply important to it as farmers and workers extracted practical benefits.

Even under ideal circumstances road building was difficult in many parts of Mexico. Heavy vegetation, harsh terrain, and high precipitation slowed construction and forced engineers to contend with the threat of washout. Across the nation uneven mountainous terrain complicated the grading process and, combined with weather, led to landslides that forced crews to clear away the debris before they could proceed. These environmental conditions took their toll on equipment and bodies, as mechanical breakdowns and physical exhaustion ensured regular turnover in the machines and personnel used to build roads.[32] Due to these challenges, government planners exhibited a willingness to use flexible definitions for the kinds of routes they built. Reminiscent of the French administrators in colonial Africa, Mexican officials in the 1920s and 1930s characterized inferior (but cheaper) dirt and gravel roads to be equal additions to asphalt-concrete routes in the country's "modern" motorway network.[33] This tactic, although politically expedient, created significant long-term conflicts as many communities later voiced anger over shoddy construction methods.

Between Nuevo León and Veracruz environmental and demographic characteristics greatly affected the organizational structure of roadwork bureaucracies and employee salaries. In Nuevo León construction efforts concentrated around Monterrey due to the city's outsized economic importance, which facilitated, from early on, a consolidated power structure that gave state government strong bargaining power with organized labor. It standardized pay rates across the state for skilled and unskilled (peon) labor and successfully resisted pressure from the Sindicato de Empleados y Obreros Constructores de Caminos (Union of Employees and Construction Workers for Roads, SEOCC), the one union that represented all road workers in Nuevo León, to more quickly raise the minimum wage.

In contrast, in Veracruz multiple cities operated their own roadbuilding organizations, often with budgets separate from the state government, constructing a statewide web of motorways that had no single point of convergence. Due to this more regional approach to road building, no coordinating structure nor single union emerged to successfully standardize pay rates, especially for unskilled workers. Although the state Departamento de Comunicaciones y Obras Públicas (Department of Communications and Public Works, DCOP) and Junta Central de Caminos de Veracruz (Central Road Board of Veracruz, JCCV) attempted to fill this void, they were only partially successful in doing so for much of this history. State officials were forced to negotiate regionally to recruit workers, and the cost of labor in the 1930s and 1940s reflected this issue, as the minimum wage paid on road-building projects varied widely across Veracruz. At the same time, given the state's geography and population distribution, the data suggest that *peones* in areas that paid less for the same work could not easily move to (or may not even have known about) higher wages in other parts of the state. The following chapters further examine these dynamics to better understand how road-building efforts developed over time, reflecting (and responding to) important historical changes under way in Mexico.[34]

Organization and Scope

This book begins in the early 1920s with President Álvaro Obregón's program for new motorways, Mexico's first attempt to implement a comprehensive road-building program. It examines this issue alongside regional lawsuits against U.S. oil companies that unsuccessfully pressed for open access to industry-built highways in Veracruz. These initial failures deeply influenced public authorities serving under President Plutarco Elías Calles, who made key legal and budgetary reforms that led to sustained road-building efforts after 1925.

Chapters 1 and 2 examine the legal and political battles that occurred in the creation of Mexico's road-building bureaucracy. They cover the day-to-day operations of the new federal agency and its state-level counterparts in Nuevo León and Veracruz. These chapters narrate the budgetary and policy shifts that occurred in the 1920s and early 1930s in response to changing economic factors, including the impact of the Great Depression. During this time the national government increasingly ceded decision making over road construction efforts to the states in return for a new cost-sharing program that reduced federal contributions to construction efforts. These chapters also explore questions related to who could use public and private motorways, issues of corruption in road building, and the rise of new bus and taxi cooperatives.

Chapter 3 covers Mexico's road-building program under President Lázaro Cárdenas. I argue that Cárdenas pursued pragmatic strategies to support construction efforts, in particular expanding the use of road bonds that created an important new role for private-sector (and later foreign) investment.[35] The courts played a key role during his *sexenio*, interpreting a constitutional right for citizens to have unrestricted access to roads. At the state level divergent trends were under way: in Nuevo León road workers formed a labor union to fight for higher wages and more social and medical benefits, while Veracruz's workforce relied on a decentralized network of agrarian and social organizations that had been actively lobbying on their behalf for the better part of a decade. This chapter continues an examination of bus services in both states, considering how operations expanded and the role that road-building agencies played in regulating these ventures.

Chapter 4 studies how road-building policy served as a key discursive space for the United States and Mexico to resolve differences over nationalization of the oil industry. President Manuel Ávila Camacho greatly expanded the use of road bonds to finance construction efforts and accepted up to 30 million dollars in investment from the U.S. Export-Import Bank. He also allowed his hot-tempered brother, Maximino, who had a long history of cronyism and political venality, to take charge of the Secretaría de Comunicaciones y Obras Públicas (Secretariat of Communications and Public Works, SCOP). Amid wartime furor governors and other officials increasingly justified new motorways with a militarized discourse once Mexico appeared ready to join the Allied forces in the Second World War. They talked about troop readiness and resource mobilization for the war effort when making the case for federal support for road-building efforts. In addition this chapter examines the regional impact of wartime shortages as interagency competition occurred over gasoline reserves, vigorous black markets emerged for rubber tires and spare automobile parts, and state road-building bureaucrats made desperate attempts to independently secure much-needed resources to ensure that agency operations continued.

My study concludes in the 1950s as Miguel Alemán's presidency settled the question of how Mexico would finance and build its motorways. Whereas Veracruz had offered a radical, populist model that empowered agrarian communities as a cornerstone of regional road-building efforts in the 1920s and 1930s, national consensus ultimately solidified around policies Nuevo León's political and commercial elites had championed for decades. Their support of professional labor brigades, private contractors, and private investment established parameters for motorway development that Alemán's and subsequent presidential administrations continued into the next century.

Chapter 5 emphasizes this growing national consensus for road building that endorsed the use of private investment to finance efforts. It assesses Alemán's use of populist political language tied to motorway infrastructure that ultimately favored elite commercial interests amid growing regional criticism over corrupt practices in road building. By the late 1940s the state road-building bureaucracies increasingly resembled one another, at least in form if not entirely in function. Both states had adopted the use of professionalized labor brigades and private contractors and relied on bonds to finance construction. Regional power brokers and local caciques continued to fight vigorously for control of local funding mechanisms. This chapter also scrutinizes key legal shifts in the late 1940s and early 1950s in how Mexico's Supreme Court reinterpreted the laws that governed road access and land expropriation for the construction of new motorways.

Chapter 6 considers the legacy of road building in Mexico and how these efforts were closely tied to state and local politics. It examines the propaganda and marketing of motor tourism, noting how travel maps, decorated with images of automobiles and other symbols of progress, depicted a highly attenuated idea of roads as an organized, rational system. Deconstructing the ideas these maps represented, the chapter revisits the contentious political landscape that national and state road-building agencies faced.

Ultimately road building in the three decades after the armed phase of the Mexican Revolution reflected the larger social, political, legal, economic, and cultural debates occurring in the country. Elite and nonelite citizens participated in this project, transforming everyday mobility with the creation of a road system with transnational, national, state, and local implications. Chapter 6 ends with a reflection of how succeeding generations of Mexican leaders in the late twentieth and early twenty-first century followed the model for motorway development set during this earlier period, using debt spending and populist rhetoric to finance and promote transportation infrastructure intended to serve as monuments to their political legacy.

In many respects the engineers and labor brigades tasked with this work achieved many of the goals that national and state leaders had envisioned in road building. Despite these achievements, however, road building remained an incomplete undertaking, as many other promises of economic improvement went unfulfilled, especially in rural areas. Over time some citizens became deeply conflicted about the program of progress that new highways and *caminos vecinales* embodied, questioning whether greater integration with U.S.-led capitalist globalization had been the correct course for Mexico to pursue after all.

1

"A Good Road... Brings Life to All of the Towns It Passes"

The Fight for a National and Public Road-Building Program

In the spring of 1921 Francisco Malpica Silva, the general manager of Veracruz's *El Dictamen*, joined a staff reporter to document his experience traveling in a motor vehicle from the port of Veracruz to the state capital, Xalapa. The main road between the cities was an old earthen path that snaked across roughly 100 kilometers of mountainous terrain and forest from the humid shores of the Gulf coast. It was a notorious example of the toll weather and institutional neglect often took on even economically important routes. Built by the Spanish, the road had helped transform Xalapa into an integral trading hub by the eighteenth century; however, state authorities largely abandoned it following the introduction of railroads to the region after 1880. The next forty years saw the road succumb to the effects of entropy as harsh weather conditions gradually made it difficult to traverse.[1] Malpica Silva's travelogue describes the driving conditions: "The car was only able to accomplish the trip, overcoming tremendous difficulties, because oxen and mules were used to help the vehicle navigate through some of the most treacherous parts of the route. Nevertheless, the amount of public enthusiasm awakened by this trip has helped to plant the first seeds in favor of automobile-centered development."[2]

This journey from the port to Xalapa was not the only trip of its kind undertaken to highlight the problems of motor mobility in revolutionary Mexico. In 1917 the national daily *Excélsior* dispatched a reporter to travel from the federal capital to Veracruz with a handful of Mexican and U.S. automobile enthusiasts. After several days of driving, the group finally arrived at the coast and hailed their excursion as a "major accomplishment" for motor travel in Mexico since their car had not broken down beyond repair while traversing the difficult terrain.[3]

A mixture of optimism and frustration marked national and state newspaper coverage on road conditions in Mexico. While travelogues emphasized treacherous driving on existing routes unfit for motor travel and editorials decried the lack of asphalt-concrete motorways, official announcements for future road construction made front-page headlines with favorable coverage. In 1917 *Excélsior* reported, "Soon a marvelous network of motor highways will surge across the country." The newspaper described twenty-seven early stage construction and repair projects under way across seven different states that promised to "invade the land" with new roads. Likewise in 1919 *El Universal* described plans for a "colossal highway" from Tuxpan, Veracruz, to Puebla that would benefit regional oil-drilling operations. Other articles suggested that Mexico needed to follow the example of "civilized countries," where good roads led to increases in automobility and economic growth.[4]

When delays or cancellations affected these projects, newspapers called on the federal government to act. In 1920 *Excelsior* emerged as one of the most vocal critics, arguing that existing road-building efforts were piecemeal and haphazard. It stated that regional caciques, wealthy citizens, and private oil companies financed distinct motorway ventures, but all of this work occurred without a coherent national strategy. The newspaper reported that the 1917 plan for road construction, which the Secretaría de Comunicaciones y Obras Públicas (Secretariat of Communications and Public Works, SCOP) devised amid considerable media fanfare, had since languished with little official attention paid to it. Moreover in April an essay by Mexico City's Automobile Club, a group of elite motoring enthusiasts that included *Excélsior*'s founder, Rafael Alducin, urged the federal government to pass comprehensive legislation to address the issue of bad roads.[5]

In response to popular pressure and public calls for action, President Alvaro Obregón finally outlined a plan to invest in the repair of existing roads and the construction of much needed asphalt-concrete routes. During his 1921 state of the union address, he framed road building as part of a larger program of economic and infrastructural reform. The president said that SCOP had earmarked more than 680,000 pesos in its annual budget for national road-building efforts. Obregón also praised the foreign-owned oil companies that operated in Veracruz and Tamaulipas for building motor roads from interior oil-producing regions in the Huasteca Veracruzana to the ports of Tuxpan, Veracruz, and Tampico, Tamaulipas.[6]

Obregón's rise to power was an important national victory for northwestern elites. He had joined with fellow Sonorans Plutarco Elías Calles and Adolfo de la Huerta to oust Venustiano Carranza and establish a new political bloc that hoped to rule Mexico. Although they had viewed Carranza as a rival and an enemy, they held many beliefs about modernization in common. They agreed on the construction of a new state along capitalist and nationalistic lines, and they saw road-building efforts as a means to support private business interests and foster economic growth at a time when foreign bankers were eager for Mexico to resume payments of its national debt. Moreover, as Luis Anaya Merchant has noted, it may come as little surprise that the revolutionary caudillos who most fervently endorsed road building and automobility hailed from northern Mexico. Given their geographic proximity to the United States, Obregón, Calles, and Carranza had likely been aware of the large-scale infrastructure projects across the border, which may have inspired them to carry out similar programs in Mexico.[7]

In 1921 the federal government under Obregón prioritized building roads that improved access to the country's coastal ports and overland borders. This strategy included motorways from Mexico City to the ports of Acapulco and Veracruz, as well as a major road to the border from Enseñada to Tijuana, Baja California.[8] One of the most ambitious projects that emerged in this period was the plan for a trinational road that united Canada, the United States, and Mexico called the Meridian Highway. Automobile enthusiasts and pro-business associations in the U.S. Midwest had developed the idea, inspired by another continent-spanning project: the Lincoln Highway, from New York to California. They envisioned a route that began in Winnipeg, Manitoba, crossed the border near Pembina, North Dakota, entered Mexico at Nuevo Laredo, and then continued to Mexico City. By the end of 1920 U.S. construction crews had already completed the route to the Mexican border, and U.S. representatives of the International Meridian Road Association met with officials from Obregón's administration, the state governors of Nuevo León and Tamaulipas, and local pro-business associations to discuss the next phase of the highway.[9]

The Chambers of Commerce in Laredo and Monterrey played an important role in generating early cross-border collaboration on the Meridian project. In March 1921 a group of Texans involved in the undertaking offered financial data and survey maps to support any feasibility studies their Nuevoleonense counterparts needed to conduct. By the following month Governor Juan M. Garcia signaled his support and instructed the state's Department of Justice, Public Education, and Development to begin working with federal, local, and foreign groups already involved in construction planning.[10] Supporters in Nuevo León saw the Meridian Highway as a potential boon to regional trade, especially the tourism industry, as automobile clubs from the United States organized driving tours into Mexico. The initial draft of the road project envisioned a route south from Nuevo Laredo to Monterrey and then westward to connect to the picturesque city of Saltillo, Coahuila, before continuing to Mexico City.[11]

Across Mexico at this time regional civic improvement committees formed to organize fundraising campaigns and recruit labor in support of motorway infrastructure. For example, in Tuxpan, Veracruz, in 1926 the inaugural issue of the *Boletín de Mejoras Materiales de Tuxpan* reported that the local ladies' guild and five private citizens donated 4,937 pesos to coat the city's main road with asphalt and concrete, while the state government contributed 500 pesos to the effort. The newspaper also noted that local oil company managers had offered to lend construction equipment, while other residents volunteered their labor. The bulletin described the work as a broad, grassroots collaboration that involved "all social classes, workers, capitalists (even the oil companies have offered to give all of the asphalt needed), men and woman who desire to beautify Tuxpan and help it advance on the road of progress."[12] Similarly in 1928 in Paso del Macho, Veracruz, citizens formed a local road board, which collected donations that ranged from five to 300 pesos. They held dances, festivals, and other civic events to help boost income when construction funds were low. Over two years private donations and public fundraisers accumulated more than 3,200 pesos, which the committee spent on road and bridgework.[13]

Newspapers touted road building as necessary to economic development in Mexico. *Dictamen's* Malpica Silva and *Excélsior's* Alducin were not alone in their enthusiastic support for new construction efforts. The cofounder of Monterrey's *El Porvenir*, Jesus Cantu Leal, and the editor of *El Norte*, Rodolfo Junco de la Vega Voigt, also promoted infrastructure development as critical to the nation's future.[14] In 1926, in an editorial entitled "We Need Highways," *El Porvenir* described new roads as an integral component of economic growth, arguing, "Our country has not been able to appreciate, in all of its magnitude, the benefit of highways, because we have been unable to enjoy these routes of transportation. In modern countries, roads have long received greater impetus, above all other improvements, to the collective interest, because national prosperity, the economies of the communities, have been stimulated in a decisive manner."[15] The editorial continued by citing the commercial benefits that railroads had had on the development of Monterrey. Drawing on the history of the city's industrialization and market growth, it characterized motor highways as the next chapter in this story. Not only would these routes spur greater industrial activity, but they would also bring new streams of income from foreign tourists eager to tour Nuevo León's countryside. The essay ended by imploring the government "to take notice of these needs in order to combine market energies with the determination necessary" to carry out road construction.[16] Journalists repeatedly made the case that good roads were necessary for broad economic growth. Their stories included images that depicted all of the stages of road building: tractors plowed land for a path, completed macadam and asphalt lanes extended to the horizon, and drivers in late-model automobiles were shown taking new routes.[17]

State and federal teachers joined in efforts to lobby government officials and the public in support of new motorways. Mary Kay Vaughan has written that the Secretaria de Educación Pública (Ministry of Public Education, SEP) and local educators played a central role in promoting progressive values, utilizing rural schools as a locus for spreading a message designed to turn peasants "into patriotic, scientifically informed commercial producers."[18] The campaign for road building was closely tied to this project of modernization. In November 1921 the teacher and road-building advocate Carlos Barrios wrote a pamphlet to appeal to "lovers of progress" (*amantes del progreso*) in favor of a highway from Zaragoza, Puebla, to Tecolutla, Veracruz. Although a teacher's salary made his own car ownership unlikely, Barrios romanticized the transformational effects of motor travel for the country as a whole; he imagined a future countryside crisscrossed by automobiles that generated economic growth from tourism that would help finance public education. He saw new roads replacing the crude footpaths that existed in much of rural Mexico, reducing regional isolation by making it easier for cars and buses to reach once remote communities.[19] In the pamphlet, which Barrios distributed in Puebla and Veracruz, and a copy of which he sent to President Obregón, he made his case for the highway:

> The beautiful idea that many of us—lovers of progress—have had, is lofty and great . . . an ample route through our lovely, rich forests; it will be the artery that carries life to all of the towns it passes. . . . The Zaragoza-Tecolutla highway will be built by the people for their own benefit. . . . *All conscientious men who love our land will participate, ensuring that every town where the road passes will be able to generate funds from regional traffic dedicated to conserve the route and to support public education.* Beautiful idea! Onward people! Help us, so that we may achieve our well-being![20]

Efraín F. Bonilla, another educator from Zacapoaxtla, joined the chorus in favor of the proposed road. On November 15 he published a flyer entitled "Proof of Our True Progress Is the Highway They

Will Build from Zaragoza to Tecolutla." He reiterated many of Barrios's themes, arguing that "progress . . . travels all across the world as a herald, its clarion call announces the evolution and revolution of our traditions." Bonilla identified road building as critical to this process, but warned of "retrogrades" who "shout spurious claims, opposing the great opening of the Zaragoza-Tecolutla Highway by whatever means at their disposal." He used ugly stereotypes of rural people to describe these opponents, calling them "illiterate farmers who oppose [this work] to the point of taking up arms" and "reveal ignorance of electricity, fearing it is the work of the devil."[21]

Bonilla pointed to "progress" as an irresistible force that "surges forward and woe to those that try to weight it down." On the one hand, he highlighted the transformative effects of building the highway, writing, "Very soon we will have the satisfaction of seeing our wild mountains traversed by automobiles. Then, this rich land, once poorly exploited due to the lack of a good road network, will enjoy an era of great prosperity." On the other hand, Bonilla portrayed opponents to the project as closed-minded, trapped by superstitious beliefs that saw technology as witchcraft. He stressed that these people, all of whom remain unnamed and ultimately faceless within his account, were willing to employ violence in their opposition to progress. At the end of the piece he reveals this local hostility to be short-sighted and perhaps hypocritical: "It will not be surprising to see some of these same individuals who had once been obstructionists, become some of the first ones to take advantage of the benefits of this important public work."

Given the nature of Barrios's and Bonilla's criticisms, we can speculate that some of the people who resisted the road project may have seen it as a threat to their existing livelihoods and local culture. They may have viewed motor roads as another example of outside exploitation of their communities by the government or private enterprise. In studies on railroad construction, John Coatsworth has shown that reduced transportation costs amplified the power of landed elites who used political and economic clout to consolidate their hold on rural land, enlarging the size of their estates at the expense of local campesinos.[22] It is not unreasonable to conclude that rural communities with memories of the complex challenges railroads brought may have also been skeptical of newer proposals for motor roads.

Challenges to Obregón's Roads Plan

The creation of cost-effective and weather-resistant routes was an important priority for Obregón's government, but high-profile engineering failures endangered the program. Carrying out roadwork was a painstaking process that required precise engineering specifications. For example, after road crews cleared away vegetation and cut an earthen lane, all excavated materials had to be removed and the lane swept clean. Next came the first layer of macadam rock, and any irregularities discovered in the evenness of the plane had to be addressed immediately. Afterward a steamroller sealed this initial tier. The asphalt-concrete coating, which consisted of a mixture of petroleum, sulfur, rock, and sand, was applied to the road at a temperature between 232 and 260 degrees Celsius. All materials had to remain absolutely dry and conform to specifications in size and weight to ensure that the concrete blend remained free of impurities. The introduction of water or deviations in the material proportions of gravel would weaken the road's structural integrity and lead to premature deterioration.[23]

One of the persistent problems that SCOP faced in the early 1920s was finding qualified specialists from within its ranks to execute a nationwide road-building program. During the 1910s foreign oil companies had built most of the best asphalt-coated motor roads in Mexico. Much of this expertise remained in the private sector and largely out of reach to the government. By the 1922 state of the union address, Obregón was forced to admit that a dearth of experienced civil engineers had led to considerable delays. In one instance crews working on the highway from Tijuana to Enseñada had failed to correctly grade and surface the route, which resulted in its complete loss during heavy rains, costing the government tens of thousands of pesos in wasted labor hours and materials.[24]

The national government also faced budgetary and diplomatic crises stemming from the debt Mexico had incurred with foreign creditors during the revolution. In October 1921 Thomas W. Lamont, a

high-ranking official with J. P. Morgan and Company, traveled to
Mexico City as a special representative of the International Committee of Foreign Bondholders, tasked with negotiating renewed bond
payments with Mexico that had lapsed due to the previous decade's
political and military conflicts. President Warren G. Harding heightened the pressure on Obregón's administration by tying Washington's diplomatic recognition of the new government in Mexico to
whether it successfully reached a deal with Lamont.[25]

The following months witnessed a series of difficult negotiations.
In June 1922 the agreement, known as the De la Huerta–Lamont
Treaty, was finally reached between the foreign bondholders, represented by Lamont, and the Mexican government, represented by
Finance Secretary Adolfo de la Huerta. The treaty stated that Mexico would set aside 15 million dollars in 1923 for a new fund to repay
the debt, followed by an additional 2.5 million annually.[26] While the
International Committee of Foreign Bondholders had managed to
force Mexico to levy a tax on oil exports and railroad operations,
these monies were not enough to cover the stipulated amounts.
Expecting tough repayment terms, by April 1922 de la Huerta had
already instructed SCOP to suspend most road-building projects and
to set that money aside pending the final agreement reached over
national debt payments.[27]

The repercussions were quickly felt across Mexico as the government ordered construction crews to simply abandon building materials and equipment at job sites.[28] Although the Zaragoza-Tecolutla
highway was one of a handful of projects reportedly spared the new
austerity measures, the national government failed to make timely
payments to cover the costs of construction. This caused considerable frustration among regional road-building representatives, who
finally wrote Obregón in the summer and fall of 1922, desperate
for help. In September the municipal council of Cuetzalan, Puebla,
summed up the larger regional challenge: "The towns across the
sierra have not given up on building the road . . . but our individual
labor is not enough. WE LACK MONEY and that is the reason we
come to you, Mr. President . . . to ask for some small financial aid."
Ultimately the national government acquiesced, sending out pay-

ment of the final 5,000-peso road subsidy, but only months after the original deadline had elapsed.[29]

One of the highest profile roads to fall victim to the new austerity measures was the Meridian Highway. On 4 April 1922, in a brief paragraph buried near the bottom of the third page of its second section for the day, *El Universal* reported that SCOP had canceled plans to build the full Meridian Highway due to budgetary constraints and would proceed only on a significantly truncated route that connected Monterrey and Saltillo to Laredo, Texas.[30] In May, President Obregón promised to give Nuevo León 21,000 pesos in monthly installments of 3,000 pesos to build this shorter road; nevertheless federal officials later suspended payment. Instead they offered the state limited funds for road maintenance, but even this support evaporated by the end of the year. Then, in 1923, as a stopgap measure, Nuevo León's legislature allocated 6,000 pesos of the state budget to pay for critical road repairs.[31]

National and state newspapers strongly criticized the national government's severe cost-saving measures. *Excélsior* predicted, "The government will experience a heavy and lasting loss. Abandoning all of the highways of the Republic, and with no one to attend to them, in less than two months they will be . . . heavily damaged, which will be very costly to repair." Other newspapers were much less restrained in their disapproval of this policy. Malpica Silva's *El Dictamen* decried austerity's regional impact: "Motor highways have now become the lifeblood of nations. . . . It is inconceivable that the port of Veracruz—in this day and age—finds itself disconnected from all of the important populations of the country, lacking even a modest road to go from here to the state capital."[32] *El Universal* had even sharper words, sarcastically condemning perennial budget challenges facing the national state and claiming that public officials had chosen to abandon economic growth:

> Reprehensibly, in Mexico, on the rare occasions that it wants to develop sound policy, it is tripped up by an eternal difficulty: the lack of money. . . . It is a real shame, because the expansion and improvement of our network of public roads is the secret to prosperity. Even

less still do we want to talk about the importance of tourism to our Mexico—which would have translated into public wealth and the arrival of civilized elements—because, disgracefully, tourism not only needs good roads, but also the absolute security and guarantees for order that we are still so far from being able to offer.[33]

State and local politicians responded to the erosion of federal support for new roads by proposing limited, regionally focused initiatives. In Veracruz in 1922, as the De la Huerta–Lamont Treaty went into effect, the state's radical governor, Adalberto Tejeda Olivares, intervened. Born in Chicontepec, Veracruz, and trained as a civil engineer with experience as a state land surveyor, Tejeda was very familiar with rural politics and supported transportation infrastructure development to improve the material conditions of local communities. He encouraged agrarian organizers to act and successfully formed a powerful social and political movement that increased his national profile, which years later backed his failed bid for the presidency.[34]

In September he unveiled a new labor program designed to strengthen the power of statewide workers' cooperatives. As part of this plan, he proposed to continue rebuilding and expanding the road network in Veracruz through local coordination with municipal leaders to organize financing drives. The work began at a small scale as public officials in towns and cities across Veracruz focused on making improvements to urban streets and repairing and building shorter regional roads suitable for motor vehicle travel. Pro-business associations and automobile enthusiasts also redoubled fundraising and planning efforts. Xalapa's Chamber of Commerce hired day laborers to clear portions of the road to Veracruz port and make structural improvements to the route.[35]

Oil Roads, Public Access, and the Courts

As state governments coped with the collapse of Obregón's road-building program, a contentious debate erupted in northern Veracruz over whether roads built by private companies should be opened to unrestricted public access. For the proponents of expanding Mexico's road network, construction efforts were only one aspect of a broader two-pronged approach that included concerted challenges to private ownership of regional motorway infrastructure. Local communities used the courts and sympathetic state officials to press for policy changes in favor of greater public access to existing asphalt-concrete routes.

Myrna Santiago has noted that in the 1910s and 1920s petroleum firms paid for and built many of the most advanced roads in Mexico to facilitate their extractive operations. Oil companies and financiers gained inroads into northern Veracruz as *hacendados* (estate owners) sold tracts of land to speculators. For example, in 1916 in Mexico City lawyers representing Antonio Alvarez and his wife, Doña Francisca Soto de Alvarez, signed an agreement with Joaquin Luis Garcia, Jose Maria Raz, and Manuel Riveroll, local businessmen from Tuxpan and Poza Rica. The contract ceded three lots the Alvarezes owned near Tihuatlán, Veracruz, and gave the investors rights to conduct oil and gas exploration and to build private roads and other infrastructure to exploit these deposits.[36]

Although some companies allowed limited public access, they often placed burdensome restrictions on local residents to use private roads, citing the need to protect their assets against competitors. The existing legal structure, which the oil companies used to their advantage, rested on a statute passed by the federal government in 1842 that identified certain "roads of the republic" (*caminos de la república*).[37] This classification covered roads built by the government that linked Mexico City to the ports of Veracruz and Acapulco, traveled between Mexico City and state capitals, or linked state capitals with "interior communities" (*comunidades interiores*). The statute, however, specifically exempted "roads that only travel to haciendas or ranches," labeling these routes private.[38] Oil company lawyers used this legal ambiguity to argue that since their clients financed and built the roads, they were within their legal rights to restrict public access as they saw fit.[39]

Beginning in the early 1920s citizens living in the Huasteca Veracruzana organized against the oil industry, filing writs of *amparo* against policies that infringed on their constitutional rights to freedom of movement. They argued that they needed access to these privately built roads because many of the alternate routes available to the public were dirt and gravel footpaths, which became unusable in bad weather. The lawsuits that emerged at this time illustrate pragmatic grassroots mobilization of many rural citizens, evincing a sense of national solidarity against foreign influence that presaged calls for nationalization of the oil industry in the 1930s. Heather Fowler Salamini's work on agrarian radicalism in Veracruz speaks to the alliances that campesinos, labor groups, and rural communities formed with state government officials against the private sector. She finds that political leaders in the 1920s "capitalized on the frustrations of the peasantry" in order to build power bases strong enough to challenge entrenched economic interests opposed to nationalist reforms.[40]

Moreover the popular movement to guarantee public access to private roads occurred during a time when the government under Obregón began to emphasize land redistribution to campesinos. As John Hart notes, many citizens who were excluded from the economic benefits that had accumulated over previous decades saw foreigners as exploiters. This population strongly supported national officials who seized properties owned by U.S. Americans and wealthy Mexicans, dividing the land and giving portions of it to rural farmers.[41]

In 1922 the *municipio* of Tepetzintla brought a writ of *amparo* against the U.S.-owned Huasteca Petroleum Company (HPC), challenging its practice of limiting regional access to private infrastructure as illegal. HPC was one of the biggest firms operating in the region, with land concessions given to it by Porfirio Díaz dating to 1902. The company boasted some of the best roads and had long allowed the public access to these routes, though it imposed certain guidelines to their use. Local residents were required to carry a special passport that proved they lived in the area in order to cross the manned barricades the company erected on its land. Adding to the controversy for many campesinos, HPC used the reviled *guardias blancas*, right-wing paramilitary groups notorious for attacks against rural activists, to police its private motorways.[42]

In the claim against Huasteca Petroleum, E. Reyes, municipal president of Tepetzintla, asserted that the requirement to carry special identification papers violated his constituents' constitutional rights.[43] This view was supported by Joaquín Cobos and twenty other residents of Juan Felipe, a town in Tepetzintla, who stated their opposition to Huasteca Petroleum in a letter to the governor:

> For matters related to our personal interests and to our families' we have encountered difficulties due to the Huasteca Petroleum Company, owner of the lands of Cerro Azul in this municipality, which has set up gates on the road . . . blocking passage to everyone who does not always present the appropriate identification issued by Mr. Ventura Calderón, head of the Department of Land Management for the company. The road in question has been recognized by all of the residents of the neighboring towns as a national one, not a private road as the company claims. This procedure [by the company] is illegal and unjust, because it contravenes the statute of Article 11 of the Constitution of Mexico . . . which holds in its spirit that all persons have the guarantee to enter, leave, and travel in the territory of the Republic without need of a security card, passport, or other similar requirement.[44]

They called HPC's influence in the region "odious tyranny." By citing the federal Constitution, residents of Juan Felipe appealed to an authority above the regional political structure. They conceived of this problem as a challenge to their economic well-being and an affront to their sovereign rights as citizens. Cobos and the others pleaded with the state government to exercise its legal powers as enshrined in the Constitution and put an end to the company's restrictive policies.[45] Governor Tejeda agreed with Reyes and Cobos; he ordered civil defense units onto HPC property, forcing the company to open the barricades and allow unrestricted public access to its motor routes.[46]

Tejeda's order met a quick reaction from HPC's general manager, William Green, who accused Reyes and the others of corrupt deals and colluding with an industry rival, Compañía Petrolera El Agwi.[47] Green contended, "We have not blocked public traffic on any of the roads . . . but El Agwi and these men have engaged in all manner of intrigues to obtain new drilling contracts and gain access to lands we lawfully possess." He sued the state government and Tepetzintla in federal court. During the legal hearings Huasteca Petroleum revealed that Tepetzintla officials had been using cars owned by El Agwi when HPC workers stopped them at one of the company's checkpoints. Green urged the judge to throw out Tepetzintla's complaint and grant the company compensatory damages for Governor Tejeda's actions.[48]

The case addressed a key point of contention between the two sides: whether the roads in Tepetzintla were public or private. Cobos and his allies in Juan Felipe argued that "everyone in the area saw these roads as national ones, not private routes as the company contends." In contrast, Green and HPC carefully delineated the policy they used to control access to the routes. Where the roads crossed over public land, the company established no obstacles to traffic. Only on portions of the roads that traversed property Huasteca Petroleum owned did they set up the barriers. As a result, they argued, the company was fully within its right "to protect its interests" against El Agwi.[49] This distinction proved critical to the outcome. The court ruled that the company was attempting to control access to private land, thus abiding by the letter of the 1842 statutes, which gave them considerable leeway over local road access and traffic. Perhaps indicating uneasiness with HPC's case, the court also struck down its demand for punitive damages, stating that the company had failed to adequately prove that any property had been destroyed by the state's civil defense units.[50]

Both sides claimed different meanings for roads that affected local access and daily use. Regional communities and officials essentially argued for de facto recognition of roads as public as long as a sufficient number of residents saw them as such and utilized them in their everyday activities. Huasteca Petroleum viewed unrestricted access to the transportation infrastructure it built as a threat to business operations, leaving them vulnerable to competitors. An additional point speaks to El Agwi's involvement: local communities pragmatically and strategically used oil rivals in order to press for an advantage in favor of greater public access. The case represented an important test of existing law, illustrating the power that private companies could exercise over local mobility.

As the courts weighed in on these cases, the political situation in Veracruz remained tense. In January 1925, near Tantoyuca, Manuel Gómez, a community organizer and activist, was killed while traveling on an HPC-owned road. The left-wing newspaper *El Demócrata* reported that just before he died, Gómez claimed an employee of Huasteca Petroleum had been responsible. Local residents from Cerro Azul, where Gómez had been headed, protested against the *guardias blancas* that HPC employed in the area. State Deputy Isaac Velásquez called these corporate enforcers a threat to public safety, proposing that federal soldiers be called into Veracruz to take over security efforts for HPC. The company's managers reached a compromise, agreeing to regular government inspections of its security installations (although the paramilitary forces remained).[51] These measures proved to be inadequate for many people as state authorities and local activists continued to press for legal and political reforms. Modification of the 1842 road statutes was necessary if regional proponents of public motorways were to prevail against the oil industry's power. The election of a new president finally cleared the way for these critical changes.

New Policy and Legal Directions for Road Building

In the 1924 presidential elections, Plutarco Elías Calles campaigned on promises to make sweeping domestic and foreign policy reforms that included passage of new road laws to reinstate construction efforts and guarantee open access for the public. In December, upon taking office, he called for the creation of a special federal road-building organization, the Comisión Nacional de Caminos (National Road Commission), which officially operated as an agency within SCOP but included a board of directors appointed by the president that reported directly to him. One of the directives Calles gave the commission was to ensure that no foreign company dominated the country's road-building program. Unlike in the oil sector, U.S. firms that contracted with the federal government had to ensure that 100 percent of manual labor and 75 percent of administrative and engineering jobs went to Mexican citizens. The government also boasted that it reserved the right to unilaterally terminate any contracts with foreign companies that failed to abide by the terms of an agreement, exceeded their operational budget, or missed deadlines.[52]

Another key pillar of this policy platform was the new president's pledge to prohibit foreigners from directly investing in road building. Calles believed that part of the blame for Obregón's failure to successfully launch road construction and repair efforts was due to interference from foreign creditors. In response to this challenge, he ordered the Comisión Nacional to raise all funds via taxes on gasoline, tourism, and other goods and services related to motor travel. No money could come from debt issues to finance federal road projects. In May 1925, when U.S. carmakers offered to extend a 30-million-dollar loan to Mexico in support of highway construction, public officials rejected the offer, reiterating the president's position that roads the government built would be paid for only through funds generated domestically.[53]

President Calles promoted economic nationalism that reinforced state power in contract negotiations with foreign consultants. In her work on this subject, Susan Gauss quotes Manuel Gómez Morín, a lawyer and leading figure during this period in official discussions over monetary policy, who wanted foreign capital to be "the servant rather than a master of the Mexican economy."[54] The goal was not to exclude U.S. firms entirely but rather to create a contract process for construction efforts that put U.S. engineering companies in a politically subordinate position vis-à-vis the Mexican government. Continuing reliance on U.S. and other foreign technical knowledge was a practical and pragmatic decision. National officials would retain the power to sack foreign contractors who failed to meet deadlines or remain on budget. As Wendy Waters notes, foreign consultants helped the government "establish realistic priorities for road construction and to decide what equipment to purchase." The division of labor stipulated in contracts between foreign contractors and the government ensured its citizens occupied a majority of engineering and administrative tasks, gaining "practical experience necessary to direct construction themselves in the future."[55]

In the summer of 1925 the new commission signed its first road-building contract with Byrne Brothers of Chicago, intending to revive many of the construction projects suspended by the previous administration. The new gasoline tax paid for an estimated monthly operations budget of 1.6 million pesos. Federal authorities put Byrne in charge of surveying and planning construction of the Laredo–Mexico City portion of the canceled Meridian Highway—now part of the Pan-American Highway—as well as work on a new route initially from the capital to Puebla, with potential to expand it to the port of Veracruz. Although Byrne's contract initially enjoyed considerable positive coverage in the national press, the government did not hesitate to follow through on its threats against companies that failed to conform to terms the government set out. By the end of 1926, claiming that Byrne was over budget on its projects, the authorities terminated the firm's agreement, turning to other foreign contractors, including the Calhoun Highway Association of Cheraw, South Carolina, to consult on roadwork.[56]

The government's willingness under President Calles to challenge private corporate power extended to reforms of national road laws. In 1925 President Calles appointed Adalberto Tejeda—the same man who had clashed with oil companies as governor of Veracruz over public access to motorways—to head SCOP. The following year the government approved the Ley de Caminos y Puentes (Law for Roads and Bridges), a major revision of the 1842 statues, which instituted new guidelines that empowered state legislatures to declare roads to be public if they determined the routes served regional needs.[57] This decision was critical to changing the existing legal arrangement that governed road access across the nation and directly affected the situation residents of northern Veracruz had faced for many years.

For more than two decades the courts had played an ambivalent role at best in adjudicating road-related cases. Judges upheld private ownership of roads but limited punitive claims by companies against state and local groups. The 1926 law provided the legal foundation for what became a major shift in judicial rulings, transforming the nation's courts into a powerful defender of public access to all motorway infrastructure.

After returning to Veracruz for a second term as governor, in July 1930 Governor Tejeda signed new state regulations redefining roads and challenging the oil industry's control over transport infrastructure. The rules stated that any motorway that traveled between two townships or municipalities was deemed "public and free to use without any limitations placed on them beyond state-mandated tolls and transit laws." Tejeda also very narrowly defined private roads as those "which serve to give exit to a property . . . or which connect two or more contiguous private estates." This change to state law was immense: no longer could private companies simply claim ownership of a road because they funded the route and directed a portion of it over land they possessed. Instead, if the road crossed communities and did not exist entirely on private land, companies could not set up any proprietary controls.[58]

HPC countered, filling a new writ of *amparo* against the state of Veracruz, claiming the government infringed on its rights as a business and property owner. William Green told state newspapers he remained optimistic his side would prevail since the company had successfully fended off similar claims against it in the *municipios* of Tuxpan, Amatlán, Cerro Azul, Chinampa, Tepetzintla, and Temapache.[59] In October 1930 Miguel W. Guerrero, Huasteca's chief legal counsel, won an initial stay against the new law pending further review by the Third District Court in Villa Cuauhtémoc. Citing the 1922 case that ruled in favor of HPC and adding related SCOP and Supreme Court decisions from 1919 and 1921, respectively, he asked the federal judge to strike down the law as it applied to HPC. Arguing in favor of private ownership of motor roads, Guerrero stressed that this control was critical to HPC's ability to ensure the integrity of its operations: "The company I represent has built a number of roads at its own expense, after acquiring the rights and land to do so, in the northern zone of the state of Veracruz. These roads are largely used for the conservation and maintenance of [company] pipelines and are routes only very minimally used for local needs."[60]

In January 1931 the Third District Court released its decision on whether to block Tejeda's road law from being enforced against Huasteca Petroleum. Judge Manuel Correa Delgado ruled that the state had not singled out HPC in the legislation or abrogated any specific agreements with the company. Moreover he ruled that the initial stay Guerrero had obtained had been wrongfully granted. The court upheld the idea that the state legislature was within its right to redefine private roads as public if local needs demanded it. Not to be denied the control it had long enjoyed over the transport infrastructure it built, HPC appealed the decision. The company argued the case through the judicial system, taking it before the Suprema Corte de Justicia de la Nación (SCJN) in Mexico City. In the spring of 1938 the justices listened to Guerrero's arguments but sustained the district court's original ruling. They concluded definitively that the 1926 Law for Roads and Bridges superseded older statutes, granting Veracruz and other states the power to essentially nationalize private motorways.[61]

After more than fifteen years of legal wrangling, federal and state officials dealt Huasteca Petroleum a significant political and legal blow. The decision marked the end of three decades of control the firm had exercised over the roads built to facilitate its operations. In one respect these cases highlighted how growing competition within the energy sector forced companies to loosen their grip over regional infrastructure. El Agwi had played a role, helping Tepetzintla to initially challenge HPC's power. At the same time local communities and state officials made broad appeals to notions of citizenship and freedom of mobility enshrined in the federal Constitution. They argued that HPC infringed on their rights as citizens when the company erected checkpoints on crucial roads and forced residents to carry special identification. Although Green and Guerrero stressed economic arguments, Governor Tejeda and the people of Tepetzintla embraced a more expansive social definition of what constituted a public road. Not only could routes built by the state carry that recognition, but any thoroughfares deemed to serve society's interests could be reclassified as open to the public.

In fact the Supreme Court proved quite receptive to arguments against private control of motorways and also defended state road-building initiatives. The top judges ruled in favor of open road access in other major cases, throwing out suspension requests and ordering challengers to stop opposition to public construction efforts. For instance, in cases in 1931 and 1937 the Supreme Court ordered landowners to take down wire fencing erected on their properties and allow state labor brigades access to continue work on highways already under construction.[62]

These lawsuits speak to the complex and changing political landscape in 1920s and 1930s Mexico. Local groups and public officials increasingly challenged the influence that foreigners and private companies presumed to exercise over roads in their communities and jurisdictions. Following Calles's 1926 reforms, the legal system emerged as a major protector of open access to motorways and defended state road-building efforts. Ordinary citizens, working in concert with public officials, lodged effective court challenges that played corporate rivals against one another and forced private industry to give up power over local transportation infrastructure. These efforts also played an important role in how federal and state agencies articulated political power vis-à-vis the desires of wealthy foreign investors to acquire lucrative road-building contracts.

Backrooms Deals and Foreign Money: The Amasco Case

Reacting to Calles's strict rules against foreign-direct investment in road building, binational business ventures formed to channel U.S. money through Mexican subsidiaries. These new entities faced a robust bureaucratic and political process that reiterated the power of the national state and protected certain domestic commercial elites. In February 1932 the U.S.-backed investment company Constructores Financieros, S.A., incorporated Amasco International, S.A., as an engineering and consulting subsidiary to compete for road-building contracts. It named R. B. Creager, a prominent lawyer and Republican Party fundraiser from Brownsville, Texas, as president of the new company. He appointed a fellow Texan, Louis Swed, to oversee Amasco's operations from Mexico City.[63]

This venture formed as a means to circumvent existing laws that prohibited foreigners from investing in national road construction efforts.[64] U.S. citizens occupied the majority of seats on Amasco's board of directors, which also included two Mexicans, Arturo Bernal, the former chief of staff of the army under Calles, and Ocampo N. Bolaños, a sitting member of the federal Chamber of Deputies.[65] The company pursued close ties with President Pascual Ortiz Rubio, appointing him its honorary chairman, while extolling the benefits of cross-border collaboration and promising the president could take a more active role at the firm once he left office. Creager directed Bolaños to open backchannels with federal officials to obtain road-building contracts. While remaining a member of the Chamber of Deputies, Bolaños publicly stepped down from Amasco's board but retained a 5 percent stake in the company. In return he discretely lobbied Ortiz Rubio's personal secretary, Nicéforo Guerrero.

The two men regularly discussed business matters between Amasco and the government. Bolaños sought preferential tax treatment for the company he represented and asked Guerrero to convince fiscal authorities to waive import tariffs on road equipment and materials Amasco shipped to Mexico from Texas. He also delivered bids for roadwork that undercut competitors, including Constructores Anáhuac, owned by the wealthy and powerful former SCOP secretary, Gen. Juan Andreu Almazán. These offers went directly to the president's office, avoiding the official *convocatoria* process, which normally selected private vendors and contractors to work with the government.[66]

Amasco found early success in this approach, ignoring established rules and pushing ahead with insider deals. By April the company had obtained an initial agreement to pave 57 kilometers of the Pan-American Highway outside of Tamazunchale, San Luis Potosí, with indications from government officials that this work could grow to include the rest of the northbound route to Linares, Nuevo León, about 450 miles away. Company letters signaled a high degree of confidence among Amasco's directors and Bolaños, owing to what they perceived to be a lucrative inside connection to the president's office. Later that month Creager flew to Mexico City, an expensive and showy undertaking in the early 1930s, taking some of his U.S. investors to meet with Ortiz Rubio, followed by a special trip with U.S. Ambassador J. Reuben Clark to see former president Calles in Cuernavaca. In his letters to Guerrero, Bolaños described Creager as a fast-moving, well-connected businessman who counted President Herbert Hoover as a personal friend. The deputy made frequent assurances that Amasco's chief could guarantee closer ties between the two countries as long as Mexico continued to show willingness to support the company's business dealings.[67]

Amasco's penchant to pursue insider deals with the president soon met resistance with national authorities who supervised road-building efforts. At the end of April, Guerrero notified Bolaños that treasury officials had rejected the company's request to waive import tariffs on five motor vehicles and construction materials. Later Anáhuac sued Amasco, alleging corrupt backroom deals following scop's spring 1932 *convocatoria*. The agency had selected the firms Rohl and Anáhuac to work on portions of the Pan-American Highway to Linares as well as other large projects in Veracruz. Amasco had undercut the winning bids by 200,000 pesos, taking over the planned roadwork through direct appeals to the president's office.[68]

This legal fight stretched into June and ultimately involved Mexico's attorney general José Aguilar y Maya to mediate. Amasco officials had assumed that President Ortiz Rubio could personally address the matter and push a decision favorable to the company's interests. That never happened; instead Aguilar upheld scop's decision to award contracts to Rohl and Anáhuac. Miguel Acosta, head of scop, reiterated that Amasco had failed to participate in the *convocatoria* process and thus had no legal standing to appeal the ruling.

Correspondence between Guerrero and Bolaños suggested that Ortiz Rubio was either unwilling or unable to press the issue.[69] In May the president had skipped a meeting with Creager when the director returned to Mexico City, perhaps to distance himself from Amasco. Bolaños criticized the decision as a waste of Creager's valuable time, but only promises to reschedule were made. By this time the company was experiencing problems with the original deal it reached to pave portions of the Pan-American Highway near Tamazunchale. scop had yet to process the agreement the company reached with the president's office. Well into June, Bolaños had repeatedly written Guerrero to do something, highlighting the July 1 deadline to finalize the contract so Amasco could receive funding to carry out the roadwork.[70]

With only days to go before the deadline passed, SCOP's Acosta raised new concerns about the Tamazunchale deal. Bolaños and Swed appealed to Ortiz Rubio, emphasizing that without this agreement Creager could not guarantee investors' confidence in the company. By early August it was clear that the efforts to save the deal had failed. Swed wrote Ortiz Rubio a final letter, accusing Acosta of deliberately sabotaging Amasco's highway contracts. In his reply the SCOP chief defended his actions, arguing that the company had acted outside official bureaucratic channels, which called into question the legality of their contracts. With the agreements abrogated, investors withdrew support, and Amasco collapsed soon after. A few weeks later, in early September, Ortíz Rubio resigned from the presidency; later that day Calles and the Partido Nacional Revolucionario replaced him with Abelardo L. Rodríguez.[71]

Amasco's failure illustrated the complexities of navigating the federal bureaucracy and national politics even with seemingly well-connected public-private partnerships. The company was not necessarily more corrupt in its dealings than any of the other major players vying for road-building contacts. Almazán had previously served as head of SCOP while his company, Anáhuac, won lucrative deals to build motorways. Stephen Niblo, who has written on the subject of corruption, finds that it occurred as "a matter of degree. . . . The difference between sharp business practice and theft can be extraordinarily fine."[72]

The principal error that Amasco's directors made was simply to have misread the national political landscape in Mexico. They mistakenly identified Ortiz Rubio as a key interlocutor for backchannel deals and plied the leader and his cohort with favors. Ortiz Rubio may have welcomed the chance to cut deals on the side, until the arrangement ran afoul of more powerful vested interests within society. Initially it appeared that they had successfully established a lucrative relationship, quickly securing contracts and evincing confidence that Amasco could serve as a go-between for the Mexican president and U.S. investors. Only until it was too late did this strategy prove ineffective, as Ortiz Rubio's men failed to protect Amasco's interests against legal and procedural attacks.

The company clearly reached too far when it threatened Anáhuac's contracts with SCOP. By opposing Almazán's powerful economic bulwark in the government, Creager and his associates overestimated their own political clout. Ample access to U.S. cash and personal ties with Ortiz Rubio and Hoover were not enough to ensure good relationships with Mexican authorities given the existing nationalist political climate that domestic elites like Almazán used so effectively to their own advantage.[73]

Road Building and Economic Nationalism

Since the mid-1920s the Mexican government had framed road building as part of a larger nationalistic program in favor of economic modernization. After 1932 President Abelardo L. Rodríguez and his successor, Lázaro Cárdenas, reaffirmed this view, promulgating policies that reiterated the country's authority over its transportation infrastructure.[74] In 1936, as SCOP completed the Pan-American Highway from Laredo to Mexico City, nationalist writers acknowledged the boon to the economy that new roads promised but also voiced long-standing concerns about entrenched corruption. Days before the highway's inauguration, Nemesio García Naranjo, a columnist at Monterrey's *El Porvenir*, cautioned the country not to repeat past mistakes. "The Laredo-Mexico highway will bring a lot of money to the country," he wrote, "but [we] must be on guard to ensure it does not become like petroleum. . . . What has Mexico gained from the millions of gallons of oil extracted from its land? Not one cent of that scandalous boom was invested in building sustainable prosperity."[75]

The steps President Calles had taken after 1924 had already begun to respond to these concerns, although addressing entrenched corruption remained a much more challenging issue. His government had implemented policy measures and new laws to ensure the country controlled how it financed and built new highways. Not only had national authorities successfully limited the extent of foreign involvement in road building after 1925; they also stipulated that administrative, engineering, and labor jobs went principally to Mexican citizens and Mexican companies. In the eyes of many of the country's nationalists, Mexico was overcoming decades of economic and political obstacles to growth as it built new roads and highways, even as it continued to combat graft and cronyism.[76]

The policy initiatives Calles launched expanded upon his predecessor's construction agenda. He reopened work on many existing road projects but aggressively addressed the structural and logistical problems that had plagued prior construction efforts. In particular the creation of a new federal agency to coordinate road-building efforts facilitated negotiations with private contractors, leading to agreements that benefited Mexico and helped to train a new generation of civil engineers. The next chapter turns to the everyday impact of this bureaucracy as the Comisión Nacional de Caminos confronted a host of political, social, and environmental challenges that emerged once labor crews broke ground on the president's ambitious road-building agenda after 1925.

2

"Everyone Was Ready to Do Their Part"

Road Politics and State Bureaucracies Take Shape in Nuevo León and Veracruz

When Salvador Toscano first looked out across the arid and mountainous landscape of northern Mexico he may have sensed the enormity of the task ahead. In 1927 he arrived in Monterrey as the first chief engineer of the Northern Division of President Plutarco Elías Calles's National Road Commission. At the time the region, like much of the rest of the nation, contained a patchwork web of roads that varied considerably in quality and coverage. Local chambers of commerce in towns and cities across the eastern U.S.-Mexico borderlands had advocated for and attempted to finance small regional projects that fed into larger developmental ambitions. Toscano's objective in northern Mexico was to build a new two-lane, asphalt-concrete highway from the border at Laredo, Texas, to Monterrey, then continue that route on to Mexico City. In a profile the *New York Times* described him as a stoic individual, "a man who never smiles," who built roads "with the finest masonry . . . trimmed to a point where the most dilapidated flivver can negotiate without trouble." With a U.S. education and laconic, scientific professionalism, Toscano was, the paper asserted, the embodiment of a new kind of Mexico that eschewed the political chaos of the previous decade in favor of social and technological "progress."[1]

President Calles echoed these sentiments when he characterized the federal road-building program as an important part of national rejuvenation. He articulated an agenda that not only looked to construct and maintain new roads but also used the Comisión Nacional de Caminos to promote the socioeconomic benefits of infrastructure development. In 1928 he boasted that the agency had organized a new National Roads Congress, which served as a forum to bring together specialists from across the country to collaborate with federal authorities and representatives from the private sector.[2] From the mid-1920s to the early 1930s, with Calles as president and later *jefe máximo*, there existed considerable innovation and variation in the organizational structure of the bureaucracies tasked with road building. The federal government encouraged state-level adaptation through a joint-funding mechanism and little intrusive oversight.

During this time Mexico faced many of the same economic and diplomatic pressures from foreign powers that it had borne under Obregón's tenure. As a political force, *Callismo* was a continuation of the political priorities set into motion at the beginning of the 1920s. The goal remained the modernization of the country and the development of a strong national state friendly to private business interests. Calles went further than Obregón in creating the institutions necessary to execute the government's policy ambitions. He also ensured that the operations of the new road-building bureaucracy would be largely subsidized by domestic taxes on gasoline and alcohol. On the one hand, the decision to forgo solicitation of funds from foreign investors underscored Mexico's poor standing in credit markets after many tumultuous negotiations with the United States and Great Britain during the 1920s. On the other hand, it offered the government the chance to characterize road building as not only a modernizing project, but a nationalistic one.[3]

In 1930, five years after its establishment, the national commission had completed work on 1,420 kilometers of new federal highways, including the strand of road that connected Laredo to Monterrey. The remainder of the fledgling network of highways extended out from the federal capital, connecting it to Cuernavaca and Acapulco to the south, Toluca to the west, Pachuca in the north, and the city of Puebla to the east. State agencies also made significant contributions, building thousands of kilometers of state highways and local roads that opened broad access for regional bus and taxi services by the early 1930s.[4]

Amid the successful expansion of the road network, instances of corruption took on subtle forms. Reading between the lines of procurement reports and everyday correspondence within the road-building bureaucracy, one finds an undercurrent of activities that quietly took advantage of the large sums of money coming in to develop regional transportation infrastructure. Official complaints noted long delays in acquiring materials and making payroll. Local communities protested short-sighted planning efforts that used cheap, low-quality surface materials for new roads, which required frequent repairs. Licensed buses and taxis confronted "pirate" services, demanding that the police do more to prevent illegal price cuts on fares.[5]

Between 1925 and 1933 the Comisión Nacional de Caminos operated largely independently from the rest of the Secretaría de Comunicaciones y Obras Públicas (scop) and maintained close ties with the office of the president. Calles appointed a special three-member board to oversee the new agency and administer the gasoline tax that financed its budget. From Mexico City the national commission directed engineering work and played a critical political role that managed media inquiries, organized technical conferences, and supervised work with private contractors. It established a first-stage plan to build five major highway projects, including routes from the capital to Laredo, Guadalajara, Acapulco, and Veracruz as well as a separate motorway from Matamoros, Tamaulipas, to Mazatlán, Sinaloa, known as the Inter-Oceanic Highway. They estimated the total cost, representing more than a decade of construction, at 73 million pesos; the Laredo–Mexico City road, 1,240 kilometers in length, was the most expensive, representing almost a third of this budget at 20 million pesos. In contrast, the 463-kilometer highway from Veracruz port to the federal capital was priced at 8 million pesos.[6]

The national commission prepared and delivered detailed monthly progress reports on construction efforts to the president's personal secretary. It also collected data that regional project managers submitted via telegraph. The composition of the commission's field teams usually included technical crews, consisting of a senior engineer and two to four junior engineers who conducted topographical surveys ahead of a series of larger construction brigades, each of which could employ as many as 120 workers.[7]

In May 1928 an eight-page dossier prepared for President Calles revealed the activities of these brigades and associated costs as they worked on the Pan-American Highway. In Nuevo León the national commission had two technical groups and four separate work crews in operation. The report stated that they had excavated more than 10,000 cubic meters of earth, employed explosives teams to cut openings through the state's mountainous terrain, and built a series of dirt embankments to facilitate road construction with the loose material. Additional work details had installed concrete tubing and built drainage ditches to reduce flooding, while pavers surfaced 46,760 cubic meters of dirt road with gravel coating and painters applied traffic markings to the finished routes to aid motorists. The national commission estimated that it paid 2.5 pesos per cubic meter of highway built; this figure translated into an estimated monthly operating budget of more than 116,000 pesos for the Pan-American Highway's field crews. At the time, taking the commission's five major highway projects into account, the work likely cost nearly 600,000 pesos in nationwide in labor and material each month.[8]

Although well staffed in technical and labor terms, what the national commission lacked was a clear mechanism for communicating with local communities to address grievances that arose during road building. In the late 1920s it relied on a variety of informal federal and state political channels to respond to problems. In some cases citizens wrote directly to the president's office, which then forwarded the correspondence to the national commission for an answer. In other instances queries went through state governors, who in turn contacted the appropriate federal officials. For example, in 1929 residents from Ciénaga de Flores and Linares, Nuevo León, wrote to the governor about the local impact of construction efforts related to the Pan-American Highway. The former wanted assurances that the national commission would not damage nearby properties, while the latter implored the agency not to alter its original plan for the highway, fearing the newer course would cause undue harm to the town's agricultural productivity. In each of these cases the governor's office, acting as an intermediary, reported to local constituents that the federal agency had taken the matters under consideration. For Ciénaga de Flores, it promised to reimburse in full any property claims the town filed, while for Linares, it agreed to proceed with the original survey route.[9]

Far from a political leviathan, the national commission responded to local concerns, albeit in an indirect fashion, while conducting a narrow mission that focused on the construction of federal highways. Neither was it the only public entity dedicated to road building, as state-level agencies across Mexico, largely separate from the national bureaucracy, worked on a number of regional motorway projects. From its creation in 1927, Nuevo León's Comisión de Caminos del Estado (Commission for State Roads, CCE) had an average yearly operating budget of 300,000 pesos. In 1931 alone the state commission worked on half a dozen asphalt-concrete motor routes, including Monterrey-Saltillo, Linares-Galeana, and Cadereyta-Allende.[10] Field crews on these projects typically employed between seventy and one hundred individuals, the majority working as manual laborers, or *peones*, for 1 to 1.5 pesos per day, while specialized staff, including clerical personnel, mechanics, and tractor operators, earned between 3 and 9 pesos a day.[11]

Although the CCE took the lead on locally oriented road building in the state, it also outsourced projects to private contractors. In the mid-1920s General Almazán's Constructores Anáhuac provided services to road-building agencies in a number of states. In Nuevo León the CCE used Anáhuac in a support role, assigning it to build bridges, widen existing motor lanes, and conduct road repairs and maintenance. In 1928, for example, state documents indicate that the agency paid Anáhuac 75,000 pesos to build a motor bridge in Monterrey that connected the neighborhood of San Luisito, in the southern zone of Colonia Independencia, to the rest of the city. The terms of the agreement tasked the company with surveying and grading the route, as well as coordinating the logistics for materials to construct the bridge. The CCE supplied the contractor with motor vehicles, including work trucks, to aid its activities.[12] Perhaps mirroring Almazán's cozy ties with federal authorities, Anáhuac also maintained a close working relationship with Nuevo León's government. Engineers the company employed later appeared in the personnel and payroll registries of the state road-building bureaucracy in the early 1930s.[13]

The organizational structure of state-level road construction bureaucracies, and the policies they implemented, were not uniform across Mexico. In Veracruz the state government initially lacked a special road-building office and instead used its Departamento de Fomento y Agricultura (Department of Development and Agriculture, DFA) for projects. Records from the late 1920s indicate that government officials emphasized urban road improvement projects, such as the repair and asphalt coating of city streets. One of the largest undertakings the DFA managed was the 1926 construction of a new road and installation of modern sewage lines and artificial lighting in Xalapa's Carrillo Puerto neighborhood, 2 kilometers south of the city center. The plan enhanced access to intermodal transport options as well as to popular tourist destinations. It connected the railroad station at Avenida Bolivia to El Dique Street, which ran alongside a picturesque lake. The DFA spent 20,000 pesos over three months on the project and relied on a diverse set of partners: it recruited 350 day laborers, hired local private contractors for specialized services, and paid foreign-owned companies for other provisions. Corporate vendors included El Águila and Pierce Oil, as well as the Mexican subsidiaries of General Electric and Westinghouse, with shareholders from the local Chamber of Commerce, which supplied construction equipment, tools, and asphalt and additional building materials.[14]

In 1930, to better coordinate statewide road-building efforts, Veracruz officials founded a new engineering authority. To do so, the state government broke up the DFA, creating an independent state agricultural commission, then combined all of the remaining infrastructure projects into a single public works agency, the Departamento de Comunicaciones y Obras Públicas (DCOP). It grew to become a powerful regional actor in Veracruz; over the following decades DCOP built thousands of kilometers of roads, collected taxes for construction efforts, and played a critical intermediary role between the state government and local communities.[15]

Redefining Mexico's Road-Building Bureaucracy

The impact of the Great Depression took a toll on the national government's ability to spend large sums on road-building projects.[16] Federal authorities looked to develop new cost-sharing programs with the states that ceded influence to governors. In 1933 it reorganized the Comisión Nacional de Caminos as part of a larger administrative overhaul to improve federal-state cooperation and budget sharing. The new agency was named the Dirección Nacional de Caminos (National Road Directorate, DNC) and came entirely under the purview of SCOP. This restructuring also created the Juntas Locales de Caminos (Local Road Boards, JLC), individual, federally recognized state-level bodies that implemented construction projects, managed road-building personnel, and corresponded with regional groups about their needs for transport infrastructure. State governors served as presidents of the JLCs, with the power to approve budget expenditures, while SCOP appointed a senior engineer who ran the day-to-day activities of the agency and served as a liaison between federal and state bureaucracies.[17]

In the spring of 1934 the national government ratified the Law of Cooperation for Roads, which formalized many of these initiatives. Along with reorganizing the national commission, it reworked the process for road-building appropriations, requiring state governments to contribute 50 percent of funds for the operational budget. This arrangement provided incentives for both sides: first, it allowed national authorities to decrease their heavy financial commitment to road building due to economic pressure stemming from the global recession under way after 1929. Second, it delegated decision-making power for infrastructure development to the states. The JLCs served as a bureaucratic mechanism for cooperation on budgetary and technical matters between the national and state authorities. New guidelines placed state governments in charge of regional road building and set up a coordinating committee composed of the governor, SCOP, the state Chamber of Commerce, and regional transportation companies to supervise these efforts.[18] To save costs, these new road-building agencies cut employee pay, reducing the minimum wage for unskilled labor to as little as 80 cents, which affected the majority of workers. In order to afford the cost-sharing program, governors also looked to private investment and began to rely heavily on state bonds to raise funds for construction projects.[19]

Although Veracruz and Nuevo León created local road boards in 1933, the operational structure of their regional road-building bureaucracies took on different forms. In Veracruz DCOP remained the primary organization responsible for implementation of statewide road construction as the Junta Central de Caminos de Veracruz (Central Road Board of Veracruz, JCCV) struggled to settle on a clear agenda for its activities due to institutional turmoil and management turnover. Moreover, in the face of seemingly intractable logistical problems, state officials allowed the creation of numerous regional committees that carried out local roadwork. In contrast, Nuevo León reorganized the CCE from early on, transitioning everyday responsibility for road building, including management of personnel and equipment to the Junta Local de Caminos de Nuevo León (Local Road Board of Nuevo León, JLCNL). Although the name Comisión de Caminos del Estado continued to be used in a variety of capacities, describing the supervisory board that the governor ran as well as the JLCNL itself, the existing structure of the older organization was subsumed into the new agency by 1934.[20]

To further facilitate federal-state collaboration, SCOP established a Department of Cooperation within the new DNC to serve as a conduit for communication between Mexico City and the regional road-building boards. Each month the JLC managers submitted detailed payroll and expense reports to the secretariat in order to receive the federal government's contribution for roadwork. The Department of Cooperation served as a clearinghouse, disbursing the appropriate funds to state accounts at the Bank of Mexico. In Nuevo León these deposits usually occurred in increments of 7,000 to 16,000 pesos every two weeks, amounting to a monthly payment of at least 14,000 pesos from federal coffers for requests ranging from administrative and labor payroll to material and equipment expenses.[21]

The department coordinated between state-level officials and federal authorities in Mexico City on regional road-building initiatives. It also allowed federal authorities to review the progress that state road-building boards made. The department did not act as an arbiter but simply facilitated the movement of national funds to the states. Once DNC officials and the governor approved the JLC's budget, these requests were processed with little additional internal scrutiny.[22]

At this time new private contractors formed to take advantage of growing opportunities the government offered for roadwork. In January 1934 in Mexico City five businessmen founded Carreteras de México, S.A. This group included Juan R. Platt, a well-connected figure in northwestern Mexico who counted President Rodríguez as a close friend and had been a personal go-between for his financial and political interests during the 1920s. A thirty-five-year-old Jewish Russian Polish émigré, Elías Sourasky, whose family invested in railroads, banking, and textiles, controlled a 23 percent stake in the firm. Capitalized at 1.5 million pesos, Carreteras de México won contracts with the federal government to provide engineering and design support to build highways, motor bridges, and *caminos vecinales* throughout the 1930s and 1940s.[23]

Everyday Aspects of Statewide Road-Building Efforts, 1930–1935

In the late 1920s and early 1930s two major highways, the Pan-American and the Inter-Oceanic, defined much of road-building policy in Nuevo León. Monterrey's role as an industrial center, with a high concentration of state residents, made the city a crucial convergence point in the development of new roads. One of the key objectives of roadwork in Nuevo León was to extend motor route access from Monterrey to regional population centers, and in turn connect these areas to more rural and isolated areas. State authorities emphasized construction of *caminos vecinales* to the Pan-American and Inter-Oceanic Highways, because these trunk lines provided easy access to the state capital. Federal officials saw this network of asphalt-concrete highways as an important economic conduit for the rest of the nation. The routes highlighted Monterrey's importance as a transshipment point with the United States, while also boosting foreign motor tourism to northern Mexico.[24]

Similar motivations were at play in Veracruz, but road building was less concentrated around a single city or a handful of national highways. Instead, in the late 1920s and early 1930s, state officials and regional communities emphasized localized construction efforts. Informal labor crews built low-cost dirt roads from rural hamlets to neighboring small towns, and larger communities looked to improve access to the nearest midsize city. This arrangement allowed smaller but regionally important places like Córdoba, Tantoyuca, and Acayucan to emerge as junctions for local road-building agendas. Bigger or more politically powerful cities, like Veracruz and Xalapa, did not dominate construction plans in the same way that Monterrey did in Nuevo León.[25]

The bureaucratic structure of road-building agencies in Nuevo León and Veracruz revealed the negotiated character of the federal-state political and economic relationship. By reorganizing as the DNC and implementing shared funding mechanisms, these policies ceded influence to state governments. Less a process of centralization, road building at this time highlighted practices of compromise wherein federal authorities delegated aspects of the construction of national highways to regional leaders. In the face of the global economic challenges of the 1930s, national and state officials formed new partnerships, coordinating planning agendas and contending with local demands for improved regional motor transportation infrastructure.

Nuevo León

In late 1932 Nuevo León had more than 1,500 kilometers of completed asphalt and macadam roads. From the state border with Tamaulipas, the Pan-American Highway traveled 400 kilometers to Monterrey and continued southward to Allende, Montemorelos, and General Terán. It then became a macadamized-gravel path for an additional 225 kilometers to Linares, before crossing again into Tamaulipas. Two other major macadam roads left Monterrey for Saltillo, Coahuila, to the west and Cadereyta Jiménez to the east. The majority of new roadwork occurred in the eastern and southern portions of the state. The route to Cadereyta Jiménez represented the initial phase of local efforts on the Inter-Oceanic Highway with planned connections to half a dozen towns, including Los Ramones, Villa China, and General Bravo, en route to Reynosa. To the south, from Linares, survey teams marked the construction path for a state highway that added a dozen towns, including the agricultural and mining region of Dr. Arroyo, considered one of the most isolated and difficult to reach areas of Nuevo León.[26]

State legislation creating the JLCNL went into effect at the beginning of 1933, and officials tasked with setting up the new agency took over everyday management of the CCE. SCOP appointed Salvador Toscano as head of the new agency, bringing years of experience to this position thanks to his previous work as senior engineer of the Pan-American Highway in northern Mexico. The complete administrative changeover occurred gradually. For example, when ordered to identify deficiencies in transport infrastructure in the state, Toscano relied heavily on CCE technical studies to articulate the JLCNL's objectives. In August 1933 he reported that many communities still lacked access to Monterrey and the state railroad network, which hurt regional economic development: "The fundamental challenge for the state, in terms of travel, is to link, by means of modern highways, all of the *municipios* within it, so that they can be connected with the city of Monterrey. There is also an urgent need to provide an outlet to regional production, for its transport along the general thoroughfares of the nation, namely railroads and highways. Likewise, the tourism factor will undoubtedly be a source of wealth for Nuevo León." The DNC was in the process of drafting a new six-year plan for road building that accentuated state-level management of highway construction across Mexico. "The state road commission is going to carefully study the program for the six-year plan," Toscano told the governor, "with the objective of . . . reasonably considering only those roads that—having accounted for the available resources of the state—can be finished." He emphasized completion of the proposed extension of roads in central and southern Nuevo León, framing construction of these routes within a nationally oriented perspective that characterized them as imperative to progress on the Pan-American and Inter-Oceanic Highways.[27]

The JLCNL's budget was larger than its predecessor's but also reflected a general reduction in federal commitments to road building due to the global recession. Starting in 1933 the agency had 600,000 pesos annually at its disposal—double the average CCE budget. Nuevo León's contributions to the JLCNL remained on par with funding amounts it had allocated to the CCE, but the national government's portion decreased sharply. For instance, in the late 1920s the federal budget allocated 1 million pesos annually to the Pan-American Highway in Nuevo León; after 1933 the new sum it provided to the JLCNL amounted to only a third of that amount. Nevertheless SCOP did promise to fund certain special projects: it paid for 100 percent of expenses incurred in a given month above 3,000 pesos for bridge construction and asphalt paving on national highways. It also financed long-term maintenance of these routes. The majority of roadwork, however, rarely exceeded the monthly cost cap placed on the state, and when it did, the excess sums billed to SCOP remained under 1,500 pesos, allowing it to easily manage any budget overruns.[28]

In response to this arrangement the JLCNL set in place key policies that had a significant impact on state construction efforts. It prioritized all roadwork into "national" and "local" categories; the former described multistate projects like the Pan-American and Inter-Oceanic highways, which were asphalt-concrete routes once completed. For local roads, which traveled between communities within Nuevo León, the JLCNL allowed for much greater variability in surface quality. High-traffic zones, especially near Monterrey, were paved with asphalt, but in rural areas the agency relied on the construction of macadamized-gravel and simple dirt routes. This decision provided significant cost savings but sacrificed long-term durability and increased vulnerability to local environmental conditions.[29]

In early 1933 a field survey Toscano and another engineer conducted highlighted the challenges facing the agency. Not only did many parts of the state lack drivable motor routes; in some areas they found that as much as 90 percent of existing roads required immediate attention to clear debris, repair potholes, and resurface the routes. The JLCNL addressed this multifaceted problem by dedicating a quarter of its field budget to repair and resurface roads. It directed construction efforts through two statewide divisions, subdivided into seven regional brigades composed of twenty-two field crews, which employed an estimated monthly workforce of more than 1,400.

The first division operated across central Nuevo León from El Jabalí in the east to the border with Coahuila, near Saltillo. The second group worked on roads that extended from Linares across the southern *municipios* of Dr. Arroyo and Mier y Noriega, en route to Matehuala, San Luis Potosí. Most *peones* in these divisions earned 80 centavos, although more experienced ones could make up to 1.5 pesos for a day's labor. Crew supervisors and specialty workers, including machine operators, bricklayers, and carpenters, earned up to 9 pesos daily. The JLCNL paid staff engineers an average monthly salary of 308 pesos, while the chief engineer, in charge of running the division, earned 661 pesos a month.[30]

The Jabalí-Saltillo group was the smaller of the two divisions, but its field crews outspent the southern brigades by as much as 2-to-1. It had an estimated monthly operating budget of 14,000 pesos, which funded one survey team and three labor units that employed a total of 450 *peones*. The group also enjoyed access to a larger quantity of machine equipment, which included eight trucks, three tractors, and a mobile concrete-mixing unit. This division's main objective was to improve local access to Monterrey, extending the central highway network to new destinations and repaving existing macadam-gravel routes with asphalt. The zone already had a high level of private and commercial traffic; whereas less populated regions could see as few as 150 vehicles in a given month, during peak seasons, between Monterrey and Saltillo, more than 3,500 automobiles, 470 passenger buses, and 1,500 cargo trucks drove through the area every month.[31]

The Linares-Matehuala division to the south operated in a largely rural part of the state. Its duties included asphalt paving of the final stages of the Pan-American Highway to the state border with Tamaulipas, and the completion of new roads through the Sierra Madre

Oriental to San Luis Potosí. An estimated monthly operations budget of just under 30,000 pesos financed eighteen field crews, employing nine hundred *peones*. Despite utilizing significantly more manual laborers, access to technical personnel and equipment remained on par with the central division, likely as a cost-saving measure. Only four engineers ran the Linares-Matehuala brigades with fifteen supervisors and support staff, three tractors, and eight trucks. Monthly receipts indicate that the division spent the majority of its budget on clearing paths through mountainous terrain and contending with repairs to routes damaged by heavy rainfall. The majority of work in the *municipios* of Dr. Arroyo and Mier y Noriega consisted of macadamized-gravel and simple dirt thoroughfares.[32]

Road building brought once isolated rural communities greater access to public services and commercial markets. In these areas citizens had long complained about the lack of medical personnel and teachers. Although still challenging, new roads gave local residents the chance to travel more easily to outside hospitals for consultations and send their children to nearby schools. Access to regular bus service, in times of good weather on macadam and dirt routes, was equally important as it allowed farmers the ability to transport goods to regional markets and to railroad stations for long-distance shipping. By the end of 1935 the JLCNL had added more than 2,500 kilometers to the state road network, which opened more of the state to motorists.[33]

Veracruz

From its formation during Governor Tejeda's second term (1928–32) DCOP took the lead on state road-building efforts. Unlike in Nuevo León, where the JLCNL replaced the CCE, in Veracruz DCOP remained an autonomous, state-level agency throughout the 1930s and 1940s. It coordinated with SCOP and supervised local construction of federal and state roads, similar to how state JLCs functioned elsewhere. DCOP's activities reflected the governor's priorities for agrarian reform and local engagement. Much more than in Nuevo León, the agency recruited campesinos to take the lead in preparing road sites and contributing labor to construction efforts. DCOP maintained active correspondence with rural communities, working with them closely on building programs that emphasized improvements to nearby transportation infrastructure.[34]

In 1930 the agency reported progress on 1,200 kilometers of federal and state road-building projects. The most prominent of these was the local portion of the Mexico City–Veracruz highway. The route departed from the national capital, traveling southeast to Puebla before turning northward and crossing the border into Veracruz near the town of Perote. From there it continued to Xalapa and then finally to Veracruz port. At 162 kilometers in length, the Perote-Xalapa-Veracruz leg represented over 30 percent of the 463-kilometer national road from Mexico City to the Gulf coast. Responding to Tejeda's desire that new roads improve rural mobility, DCOP told the governor that the route connected over two dozen small towns and hamlets to the burgeoning highway system.[35]

Raising revenues was an important part of the process but faced resistance from those required to pay. In areas where the state government built the Perote-Xalapa-Veracruz road, it levied a 90-cent tax on property owners to help offset construction costs. In June 1930, when citizens in Banderilla, a town of 2,300 that neighbored Xalapa, complained it was too much, DCOP rejected their requests to suspend the levy, raising it to 1.3 pesos for those deemed able to pay. Following negotiations that stretched across the summer, the agency finally relented, reducing the tax burden by 75 percent and allowing affected residents to pay the remaining amount in installments over six months.[36]

Citizens also pushed back against taxes that targeted their proximity to road-building projects. In October 1930 María Luisa Esteva, a local property owner in Xalapa, named the city's municipal council, the state legislature, and the governor in a lawsuit filed in federal court, claiming a road tax violated her constitutional rights. The municipal government had levied fees on land she owned in Xalapa, directing the monies raised to pay for asphalt paving and other improvements to roads adjacent to her property. In a letter to the federal court in Xalapa, Esteva wrote, "On October 23, an employee of the municipal treasury demanded that I pay the expressed sum of

77.5 pesos for paving, as well as a twenty-two peso fine." She argued that the amount demanded of her was more than 6 percent of the value of her farmland, which violated one of the limits capping tax burdens written into state law. Veracruz officials gave her three days to pay, and when she refused, local authorities in Xalapa garnished the rents she collected on the land, leading to the lawsuit.[37]

Within days of hearing her case the judge in Xalapa ordered a seventy-two-hour freeze on any further actions taken by government authorities against Esteva. In response Governor Tejeda appealed to a higher court. On November 15 the first district judge of Veracruz, Arturo Martínez, upheld Esteva's claim in a limited ruling that ordered the government to drop further prosecution of the property owner but did not require authorities to return the embargoed funds.[38] Ultimately the case highlighted the potential for conflict over transactions related to local road-building efforts. Although lawsuits remained a common facet of contentious transport infrastructure projects, in succeeding years state officials typically avoided imposing specific duties on individual property owners. Instead Veracruz authorities increasingly relied on sales taxes on alcohol, gasoline, tourism, and related goods and services in funding future construction budgets.

In DCOP's 1930 report, roughly 1,000 kilometers of road projects it listed covered a diverse assortment of medium and small-scale construction efforts. The agency built a handful of short, direct routes between larger cities and midsize towns; one example of this was the 12-kilometer road from the state capital to the wealthy coffee-growing town of Coatepec (population 19,000 in 1930). Similar thoroughfares extended from Orizaba, Córdoba, and Veracruz port, improving access for communities in the countryside to these economically important population centers. In addition the agency built longer circuitous roads that ranged between 80 and 200 kilometers, snaking through mountainous terrain that linked midsize towns such as Tantoyuca, Huatusco, and Zongolica (each numbering around 3,500 inhabitants) and small towns of fewer than one thousand residents with a terminus city like Xalapa or Tuxpan.[39]

Rather than use formal statewide construction brigades to build many of these rural roads, the government in Veracruz relied on local

communities and "volunteer" labor. This policy divided road building into segments, where successive groups of area residents contributed their own labor to complete their local portion of a larger
camino vecinal. As Waters has noted, construction efforts utilized
the *faena*, a traditional source of communal work dating to the colonial period, providing a source of cost savings for the state government. Leaders of *ejidos* (an officially recognized, community-based
system of land management for agriculture) organized male members into informal crews that cleared away vegetation and debris,
working with traveling survey engineers who instructed them on
grading dirt roads.[40]

DCOP officials who documented these projects recorded scenes
of intensive manual labor as construction teams worked with pickaxes, shovels, and handcarts. Many farmers relied on the same tools
for roadwork that they used to cultivate nearby fields. Typically a
gasoline-powered tractor was the only heavy machinery brought in
to assist these efforts. The *caminos vecinales* built in Veracruz often
lacked any kind of weatherproofing or gravel surface. The state government eschewed more costly road building, which required additional engineering expertise and heavy machinery, in favor of quickly
widening and improving existing footpaths to facilitate motor traffic in the dry season.[41]

Of course not all campesinos were thrilled to donate unpaid labor
for road building. People abandoned worksites and refused to carry
out maintenance ordered by municipal officials. In 1930 in the northern *municipio* of Tamalín, the mayor wrote Governor Tejeda about
the lack of community support for construction efforts. He complained that much of the national highway that passed through the
region remained unfinished or in disrepair. He ordered all of the
residents to help fix this problem, but the local campesino union
refused. Instead they marched in the town, denouncing the mayor
for trying to force them to work without pay. Waters finds that Tejeda
was conflicted about the *faena*; he did not fully embrace the practice but did not explicitly reject its use by local officials. He told the
mayor of Tamalín to try to convince the recalcitrant citizens that
"they will benefit from cooperation in improving the road." Tejeda

also suggested the local committee for civil improvement be convened to see if it could find any additional funds for the roadwork.[42]

Although basic construction and repair activities could be done by "volunteer" labor, more technical undertakings required state investment and planning. Bridge construction was one of the most significant investments that public authorities made. For example, in June 1931 the communities of Texistepec and Oluta proposed a budget that included repair and expansion of the old rural road that went through their region. Both towns had populations of fewer than 2,500 inhabitants and wanted better access to the midsize city of Acayucan (population 11,811 in 1930). The improved road remained a dirt path but was enlarged to 9 meters and graded to accommodate two lanes for motor vehicles. The local road committee estimated the work to cost 6,450 pesos, with 20 percent of funds going to construction of two bridges on the 13-kilometer route. The rest went to pay wages.[43]

The construction of low-cost dirt roads had long-term consequences for regional mobility. Bad weather and daily use deteriorated the routes more quickly, and communities lobbied for state help to keep paths open. Even prominent roads to economically important areas—like the one between Orizaba and Córdoba—lacked macadam or asphalt surfaces, which decreased accessibility during months of high amounts of rainfall. "It is not uncommon for whole regions to be left impassable due to the heavy rains, which block the roads for cars and other vehicles," complained one local official from Huatusco, in central Veracruz.[44]

Citizens petitioned state officials to address the problem; their requests usually took one of two forms, depending on the size of the community. In areas with high demand for motor traffic, local residents wanted existing roads to be resurfaced with asphalt. For example, residents of Orizaba and Córdoba took this course of action in correspondence with the governor's office. In less populated and poorer areas, pro-road committees simply asked for the state to contribute the materials to make necessary repairs to caminos vecinales damaged by erosion. In Xoxocotla (population 1,483 in 1930) the president of the town council, Alfonso Romero, wrote the governor about their needs: "With all respect, we plead that you provide us with a few tools to make repairs to our local road. We have made this request before, but have received no answer, so today, we respectfully reiterate our appeal to you for twelve shovels, twelve pickaxes, and five hammers."[45]

In December 1932 Gonzalo Vázquez Vela replaced Tejeda as governor, which led to a gradual shift in the way the state organized labor for roadwork. The new governor sought to bring state road-building policy more closely in line with federal procedures and to centralize the logistical and supervisory functions of the bureaucracy within the auspices of the state government. Although he did not try to eliminate local influence entirely, Vázquez Vela did reduce the level of control that smaller rural communities had exercised on construction projects under his predecessor.

In February 1933 DCOP began construction on a new state highway that stretched 38 kilometers north from Altotonga to Tlapacoyan. The agency eschewed the *faena*, responding to federal guidelines that prohibited the use of coerced labor practices. Instead it implemented a wage labor structure, grouping personnel into a series of formal work crews. DCOP divided the Altotonga-Tlapacoyan route into three distinct parts, each measuring an average 12 kilometers, and assigned a separate brigade to the individual sections. More than eighty individuals worked on the entire highway, and the largest company consisted of more than forty. *Peones* earned 80 cents per day, working six days a week, while crew supervisors and specialized workers, including drivers, mechanics, and masons made 1.81 pesos daily, on average. DCOP's central office assigned at least two engineers to supervise the entire project.[46]

Brigades implemented a combination of asphalt and macadam surfaces to build the Altotonga-Tlapacoyan Highway. This allowed DCOP to manage expenses, while also providing greater protection to vulnerable parts of the route that received higher amounts of rainfall. Workers used dynamite to cut through mountainous terrain in the region and had access to more equipment and motor vehicles, including steamrollers, pickup trucks, and tractors. The average monthly expense for personnel was 3,061 pesos, while the cost for materials, as needed, could reach 1,100 pesos in a given request.[47]

Despite labor reforms intended to make construction efforts more efficient, the Altotonga-Tlapacoyan Highway encountered repeated delays, possibly due to a lack of communication, official incompetence, or petty corruption. Throughout 1933 Ramon Mora, the project's chief engineer, complained to DCOP bosses about the agency's repeated failure to make timely payments to cover payroll and supply costs. Weather and difficult terrain also added to the problem, causing work to progress slowly. Mora did not finish the road until well into 1934, despite local offers to collect money and provide labor, which DCOP rejected.[48]

Initial technical and budgetary problems notwithstanding, state authorities continued to move away from informal labor for road building. Although rural communities and agrarian groups still volunteered to clear and repair roads, stated policy emphasized a formal wage and labor structure for agency employees. In part this change responded to national officials who frowned upon labor arrangements viewed as "backward" due to their reliance on traditional community ties over structured wage systems. Public authorities saw the kinds of building practices involved for the construction of motor roads as essential to the process of producing a "modern" nation.[49]

Local businesses began dispatching employees to assist in roadwork. They reached agreements with the state government that set a monthly quota of employees sent to staff construction crews. Like DCOP workers, company personnel typically earned between 80 cents and 2 pesos, depending on experience and responsibilities. These employees worked alongside DCOP staff and were supervised by an agency engineer in an organizational structure that resembled the system of brigades and crews used by the JLCs.[50]

These public-private partnerships were vulnerable to economic recessions. For example, in the spring of 1933 a sugar mill in San Carlos served as a major partner on the state road that traveled through the *municipios* of Ursulo Galván and La Antigua along the Gulf Coast. In July the mill's managers wrote to the state government, informing it of the need to reduce its commitment to staff road crews due to financial pressures from weakened demand for sugar in the economy. "We are holding out for better times, but as of this moment, we are not capable of continuing to make payments," wrote Harry Skipsey, the mill's owner, to the governor. "With all of the good that I have contributed to the construction of roads—investing thousands of pesos into them—while it would be a pleasure to continue to do so, for now, I ask that you relieve me of this expense."[51]

In 1934 a similar problem emerged in Tuxpan, where the local bus drivers' union had assigned some of its members to construction crews, paying them a daily wage of 2 pesos for roadwork. Citing heavy and extensive damage that severe weather had caused to many area roads, the union complained of the economic strain placed on it from continued regular maintenance to these routes. "The roads that depart this city to the east and west are in terrible conditions for travel," explained Eulalio Gutierrez, the union's secretary general, "but they are the only routes that exist for our members to work and provide enough for their families." Gutierrez asked the governor that the state take over paying the wages of its members involved in road building and maintenance; otherwise the group would be forced to reduce its labor commitment. "The work these men will be doing, in virtue, is considered a public utility," the secretary urged. Public officials acknowledged the issues at hand and agreed to increase their budgetary commitment to offset the costs for the private companies and unions involved.[52]

Despite policy changes the new governor did not abandon the established organizational structure of Veracruz's road-building bureaucracy. In fact even as the JCCV and other road-building organizations took on new responsibilities, DCOP continued to play a central role in the technical and political aspects of road building. What did change was that the agency emphasized its position as an intermediary between the state government and many rural communities. Its officials regularly communicated with state politicians and regional caciques about planned blueprints and budgets, streamlining the application process to request building materials, and visiting field sites to address local grievances about construction efforts. Increased popular demand for motor travel, and the attendant rise in automobile and bus traffic in the 1930s, also helped to push DCOP into a new position as an arbiter of how citizens used roads.

The Rise and Regulation of New Bus and Taxi Services

New kinds of challenges required dynamic policy responses, and Veracruz's state legislators, by the early 1930s, approved new powers for municipalities to set fines and prosecute motorists. Local authorities began giving out 5-peso fines for transit infractions (with maximum penalties of 200 pesos per offense), and the monies collected went into DCOP's operational budget.[53] The agency became very active in resolving disputes between transit services. For example, in March 1933 municipal authorities from Córdoba and Amatlán, neighboring cities in central Veracruz, faced a serious quarrel between their respective transport unions that handled passenger traffic in the area. The bus driver's union in Amatlán and the Sociedad Cooperativa de Choferes de Córdoba (Drivers' Cooperative) each demanded exclusive rights to transport passengers on the road between the two cities during Amatlán's annual five-day fair in early May.[54]

The dispute had been brewing for weeks, and transit officials worried that it could deteriorate into work stoppages, threatening regional motor mobility and hurting the local economy. On Governor Vázquez Vela's request, DCOP arranged for both sides to meet behind closed doors to resolve their differences. In late April, when they convened a meeting at the office of Córdoba's municipal president to discuss a deal, Amatlán's mayor attended. By the end of the session DCOP had succeeded in striking an agreement between the rival transport associations that allowed them to temporarily share their routes during the festival, averting further conflict.[55]

Regional bus service became an important way for ordinary citizens to access the country's new highways and state roads. Round-trip tickets between Xalapa and Puebla cost 4.5 pesos, while direct service from Mexico City to the border at Nuevo Laredo was 56.80 pesos. Local carriers in major cities, like Monterrey, charged as little as 10 cents for fares on urban buses. In contrast, automobile ownership remained out of reach for most average people; a typical four-door sedan in 1933 cost as much as 3,346 pesos, roughly equivalent to six months of wages for a midlevel civil engineer. Although older, used cars sold for as little as 750 pesos, fewer than 1 percent of Mexicans in the 1930s owned some type of private automobile. As such, a large potential market existed for transportation companies to pursue, and federal officials estimated that between 1929 and 1935 more than six thousand buses operated nationwide, serving tens of thousands of passengers daily.[56]

Many regional bus services began as small operations. Ordinary people pooled resources to purchase vehicles and applied for commercial licenses, using converted trucks to transport passengers. Some of the earliest groups formed in the 1910s in Mexico City and nurtured revolutionary ideals. Gradually these kinds of cooperatives spread to other parts of the country.[57] In Orizaba, Veracruz, fifteen former textile workers and railroad mechanics each purchased trucks on credit to start a local transport co-op, the Unión Camionera de Orizaba y Anexas, offering service to the neighboring city of Mendoza. In December 1936 they wrote to the state government, using strikingly political language that captured their motives: "In order to emancipate ourselves from bourgeois control and never again be its victims, each of us acquired buses to transport passengers. These vehicles constitute our entire assets . . . representing great sacrifices due to the high cost of this business, which has forced us to often deprive our homes of necessary goods."[58]

Other organizations formed as shareholder ventures, issuing stock to members and relatives. In April 1934 the Sociedad Cooperativa de Camioneros del Servicio Urbano de Jalapa y Anexas launched in Veracruz's state capital with twenty-three founders who invested between 30 and 200 pesos. Besides operating in the city, it won a government concession to carry passengers to the neighboring city of Coatepec. Most of the members, young men in their twenties and thirties, also worked as the cooperative's drivers.[59]

Financing the operation was a family affair. Ana Maria Andrade Sanchez, a fifty-two-year-old widow, and her twenty-six-year-old daughter, Carmen Sanchez Espinosa, a married office worker, invested 200 pesos in the cooperative on behalf of Ana Maria's thirty-six-year-old son, Samuel Cordoba Sanchez. Likewise two brothers, Constantino and Neftalí Espino Zamora, and their cousin, Crisóforo Cortez Zamora, worked as drivers and provided 320 pesos in total to the association. Guillermo Nájera y Olivier, a forty-three-year-old father, gave 70 pesos for his son, eighteen-year-old driver Guillermo Nájera, who had invested 30 pesos for membership. Membership dues covered operational costs as well as insurance coverage for accidents and health care, while a general assembly met twice a year, in February and August, to vote on measures and elect the administrative council that ran the cooperative's day-to-day activities. The cooperative also reserved the right to remove members for negligence, damaged property, or acting in a way that "diminished the cooperative's prestige."[60]

Early taxi services typically consisted of one or two individuals who owned an automobile and operated fares either within a single city or between two distinct regional communities. In Veracruz DCOP processed new applications, requiring candidates to submit paperwork describing the vehicles they planned to use and their condition. Applicants also had to list where they intended to operate in order to avoid overlapping service. The head of the agency responded to these requests, granting commercial licenses and directing drivers to consult appropriate conduct and safety statues in the General Road Law. DCOP reserved the right to carry out vehicle inspections prior to granting commercial licenses. In this way the agency served as an important gatekeeper to this market, determining who could legally profit from the state's motorway infrastructure.[61]

In contrast to DCOP, the JLCNL did not carry out many of the same regulatory functions. The agency did supervise roadside spaces, issuing warnings to property owners, farmers, and others to remove obstructions if they built commercial signs, irrigation networks, or other structures too close to a motorway. The policing of bus and taxi services, however, remained the prerogative of other state and municipal authorities. Given that Monterrey contained roughly one-third of the state's population, city officials exercised considerable influence over the transportation options for many state residents. In 1936 municipal authorities launched a bevy of regulations, including a vehicle inspection program for buses and trucks in the city. They also eliminated many downtown two-way streets and gave out fines for jaywalking.[62]

The rise of bus and taxi services presented a new point of entry for ordinary citizens to participate in the social, economic, and technological forces affecting the nation. J. Brian Freeman argues that drivers in these nascent transportation cooperatives represented a "new class of workers and entrepreneurs" who, quoting the famed Mexico City writer Salvador Novo, became "the first sons of the Revolution."[63] The internal rules that governed these associations, as well as the state regulations that imposed certain obligations on them, were closely tied to notions of order and progress. Cooperatives created routine, reliable service routes in (and between) communities, and also required drivers to behave responsibly or face expulsion. State and municipal officials resolved disputes between groups, conducted vehicle inspections, and set rules that limited the number of companies operating in a given area.[64]

Transport cooperatives that won government concessions to run service routes often complained to public officials about rival, unauthorized groups entering the marketplace. These pirate taxis and buses were popular among working-class and rural residents, not only transporting passengers but also carrying livestock and other goods. Authorized transportation cooperatives demanded that state and municipal officials enforce rules to end these informal services. In response, however, public authorities appeared loath to crack down on these groups, perhaps due to limited policing resources or because they received kickbacks for not taking action. For some rural residents this type of transportation was the only affordable way to use state roads in remote areas that lacked adequate coverage from sanctioned busing companies. Unauthorized transport was popular, and tensions persisted for many years.[65]

The approved and pirate bus and taxi services that formed in the late 1920s and early 1930s supplied low-cost options for ordinary citizens to travel on new federal and state roads. Whether legal or illegal, they ensured access to motorways was not limited to people who could afford to own an automobile. Instead they widened the possibility for participation in regional mobility and contributed to the general interest for road building among many urban working-class and rural citizens. Moreover public attention led to concerted efforts to improve the kinds of roads being built in order to keep routes open for buses and commercial trucks.

Road building after 1925 marked a series of new beginnings for Mexico. The national government implemented policies that funded and built highways across the country. It also forged collaborative agreements with state road-building agencies that highlighted an important instance of federal-state cooperation and power sharing. The emergence of the Juntas Locales de Caminos enhanced state voices in the planning and construction process, in large part because they shifted the road-building bureaucracy to a regional orientation.

The differences in how these structures formed in Nuevo León and Veracruz indicated that federal authorities were amenable to a considerable amount of experimentation and adaptation at the state level in return for help in funding construction work. Enforcing transit rules and ensuring that road-building projects received financial support remained significant challenges likely due to bureaucratic inefficiency and petty corruption. The following chapter examines the evolution of the day-to-day operations at the JLCNL, DCOP, and related agencies as they coped with growing popular demands for better quality roads and addressed calls for improved treatment of workers in an age of renewed left-wing populism.

3

"So That These Problems May Be Placed in the Hands of the President"

Roads and Motor Travel under *Cardenismo*

I n the summer of 1936 Mexican officials prepared to celebrate the completion of the Pan-American Highway from Laredo to Mexico City. In Nuevo Laredo they erected a temporary pavilion draped with Mexican flags at the bridge that united the new road across the border. Delegations from the United States and Guatemala, including U.S. vice president John N. Garner, joined Mexican authorities at the site, calling the highway a "symbol of peace and prosperity." In Mexico City, President Lázaro Cárdenas hosted a formal dinner to commemorate the project for Mexican and U.S. businessmen, as well as U.S. diplomats living in the capital.[1]

As this opening of the Pan-American Highway marked the end of more than a decade of planning and construction efforts, it also represented the start of a new era of motor travel and national commercial development. The highway was a boon to Mexico's budding tourism industry; hotels, restaurants, and travel centers catering to domestic and foreign motorists opened along the route. State governors and travel agencies produced bilingual driving guides that depicted points of interest, marketing popular destinations to travelers from the United States and Canada.[2]

Construction of the Pan-American Highway had required the creation of an expansive and costly road-building bureaucracy that trained a generation of civil engineers and technicians. In support of these operations President Cárdenas enhanced policies initiated by his predecessors, most important those of Plutarco Elías Calles and Abelardo L. Rodríguez, which greatly affected how public and private institutions financed and built new roads. The impact of the Great Depression reduced traditional sources of tax revenues that federal authorities had relied on to pay for construction efforts. In response the government increasingly looked to motor travel and tourism as an engine for economic growth and a means to subsidize road bonds. First Rodríguez and later Cárdenas redirected the gasoline tax and other levies to underwrite new public debt issues for future roadwork. In doing so they also expanded the responsibilities of private contractors in this work and allowed for limited foreign investment in transportation infrastructure.

During the 1930s Cárdenas largely upheld established federal policy for road building but supported changes to financing measures and labor relationships at SCOP and state agencies. Thanks to deficit spending, national and state officials dramatically increased construction and maintenance budgets, often doubling the amounts spent in prior years. In Nuevo León and Veracruz new money, old local disputes, labor activism, and a growing role for private contractors, bus companies, and transportation cooperatives dominated the socioeconomic and political landscape. The courts too played a critical part in protecting public access to motorways and forcing landowners to cede portions of their properties to state road-building efforts. In 1938 the president's decision to nationalize the petroleum industry had an acute but ultimately short-term impact on motor mobility and road construction.

As a political project, *Cardenismo* was an important period of pragmatic reforms. Cárdenas expanded the government's focus on infrastructure and economic development with policies that appealed to working-class and rural citizens. The goal no longer was to simply create a state favorable to private business interests. Instead the president encouraged collective action, through unions and agrarian organizations, to achieve social change.[3] By increasing investment in road building Cárdenas retained good relationships with the business sector while also aiding grassroots movements that fought for greater control of their land and workplaces.[4]

National Economic and Legal Reforms

The specter of the debt crisis that had bedeviled Obregón's ambitions to build new roads and grow the economy remained a palpable concern for policymakers. As Mexico gradually amortized the debt it owed foreigners, national leaders looked for new financing options for road construction. In 1934 President Rodríguez had announced plans to initiate a limited pilot program for road bonds. He cited budget pressures due to the Great Depression, as well as Mexico's timely repayment of its foreign debt obligations, to justify the scheme. Rodríguez's initial road bond generated 12 million pesos, which he earmarked for the completion of the Pan-American Highway from Laredo to Mexico City.[5]

This debt issue set an important precedent that Cárdenas greatly expanded. In March 1937 federal authorities issued a new road bond, valued at 15 million pesos, later increasing the offering by an additional 6 million pesos. In September, during his state of the union address, Cárdenas characterized the program as a critical component for his administration's strategy to pay for new road-building efforts.[6]

The decision to issue road bonds followed two key factors that favored more public spending on infrastructure. First, in December 1936 Mexican officials concluded payment negotiations with U.S. banks, successfully reducing external debt obligations valued at 267 million dollars by 75 percent with an annual interest rate of only 1 percent.[7] Second, motor tourism had steadily risen along the U.S.-Mexico border. The *New York Times* described the crossing stations in Texas at Laredo, El Paso, and Brownsville as the fulcrum of a "gigantic funnel of the North American continent," where travelers from across the United States entered Mexico. In the summer of 1935 the American Automobile Association and a satellite office of the Dirección Nacional de Caminos in San Antonio recorded a 900 percent increase in monthly requests for cross-border motor vehicle permits.[8] In Monterrey state officials reported that highway traffic in Nuevo León was up 70 percent.[9] By 1937 U.S. motor tourism to Mexico had risen 33 percent to roughly 50,000 annual visitors, and doubled to more than 100,000, the following year.[10]

Cárdenas relied on this increased traffic to finance new road bonds, arguing that the revenues generated via the gasoline tax made the debt payments affordable. By 1940 the Mexican government had raised a total of 72 million pesos from bonds to support construction efforts. This amount included a new agreement with the U.S. Export-Import Bank to make a relatively small investment of 192,000 dollars (864,000 pesos). Later, and likely as a result of growing tension over the 1938 expropriation of the oil industry, Mexican authorities did not sign any additional investment deals with U.S. agencies. Instead they offered bonds to domestic creditors via the Bank of Mexico.[11]

The president's decision to embrace public debt as a way to subsidize road construction mirrored policy trends already under way at the state level. As early as 1935 the government of Nuevo León began talks with lending agencies, including the Compañía Mexicana de Garantías, S.A., a firm based in Mexico City with capital and reserves valued at 1 million pesos. The agreement struck with the Junta Local de Caminos de Nuevo León (JLCNL) initially provided 8,000 pesos a month to subsidize the agency's operations. This figure grew precipitously, and in the summer of 1936 the company extended Nuevo León a monthly line of credit of up to 60,000 pesos.[12]

Lenders competed with one another to extend loan offers for road construction, and close ties with public officials were important to ensure that deals remained in tact. For example, in 1938 the state treasurer who had negotiated Nuevo León's bond agreement with the Compañía Mexicana de Garantías was removed from his post. Soon afterward the firm received notice that its contract would not be renewed. In the termination letter the JLCNL stated that it had decided to begin working with Monterrey-based Compañía de Fianzas Lotonal, S.A., describing it as "an institution specially recommended by the state government to execute all necessary bond-related activities."[13]

In Veracruz as early as 1934 Governor Vázquez Vela had turned to the Banco Nacional Hipotecario Urbano y de Obras Públicas (National Bank for Urban Mortgages and Public Works, today known as Banobras) to underwrite 865,000 pesos in emergency loans for road building. State officials had run short of funds for the Perote-Xalapa-

Altotonga-Xalapa Pro-Highway Committee,
Archivo General de la Nación, Fondo Lázaro
Cárdenas, Vol 626 Exps 514.63-17-515-1-10.

(*opposite top*) Motor caravan along the Altotonga-Xalapa
road, Archivo General de la Nación, Fondo Miguel
Alemán Valdés.

(*opposite bottom*) Motor bridge and traffic, Monterrey,
Archivo General de la Nación, Colección de los Hermanos
Mayo, Expediente HMCN-1705/1-A.

(*above*) Farmers providing volunteer labor to prepare the
Tuxpan-Tihuatlán portion of the Tuxpan–Mexico City
Highway, 1935, Archivo General de la Nación, Fondo
Manuel Ávila Camacho, FMAV 188-Exp 515.1-69-515.1 150.

PEMEX Road Map of Mexico, Archivo
General de la Nación, Fondo Dirección
General de Investigaciones Políticas y
Sociales, Expediente 19.

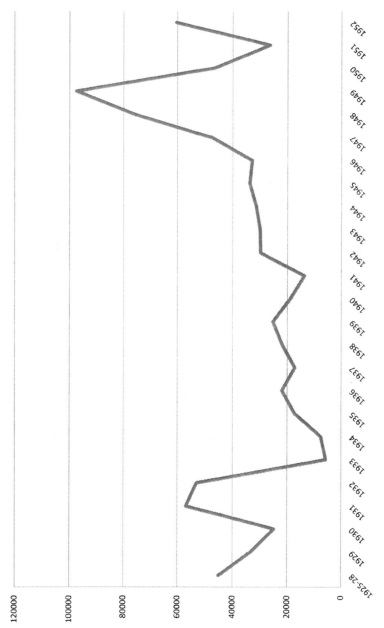

Inflation-Adjusted Federal Spending on Road Construction. INEGI, *Anuarios Estadísticos*.

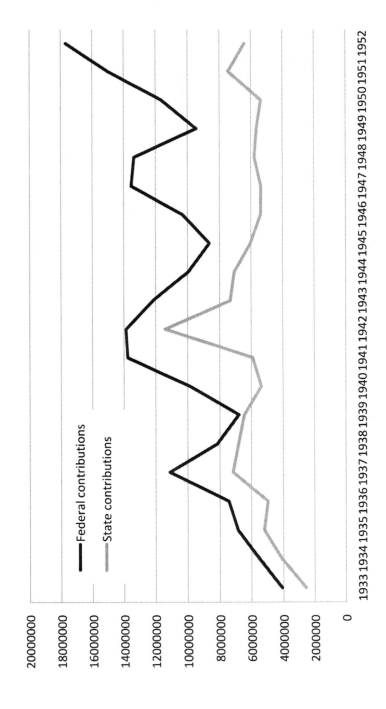

Inflation-Adjusted Federal/State Contributions to Program on Cooperation for Roads. INEGI, *Anuarios Estadísticos.*

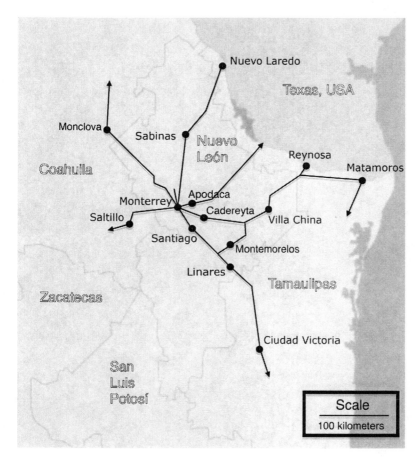

Map of Nuevo León major highways, 1950.

Map of Veracruz major highways, 1950.

Veracruz route of the highway from Mexico City to the Gulf coast. In February 1937 Governor Miguel Alemán announced another agreement with the firm, worth 1.2 million pesos, which would be repaid in installments at a fixed interest rate of 7 percent over eight years.[14]

As Cárdenas and state governments redoubled financial support for road building, the courts defended public access to motorways and protected official measures to expropriate land for construction efforts. In March 1938 the Suprema Corte de la Justicia de la Nación ruled against another major oil firm in northern Veracruz. The Compañía Internacional de Petróleo y Oleoductos (International Company for Petroleum and Pipelines, CIPO) had filed for a writ of *amparo* to protect its operations against state attempts to open access on its private roads. The plaintiff's lawyers argued that the road in question did not serve "public utility," as it connected two work camps with the company's oil fields. The judges disagreed with CIPO, citing Veracruz's new labor laws, which stated, "Big industrial owners have an obligation not to impede the free flow of traffic on any highways or roads between work areas." Thus the court not only upheld state rules but also broadened the definition of public access beyond what prior cases had already achieved.[15]

Landowners across the nation faced judicial rulings that limited their control of property. The SCJN declined to enforce writs of *amparo* that individual citizens brought against state road-building plans they claimed unjustly affected their livelihoods. In 1935 Fermín Terreros, a landowner from Jilotepec, Estado de México, fought state officials to prevent them from expropriating parts of his farm to build a *camino vecinal*. Terreros feared that if the authorities went forward with their plan, it would interrupt the planting season for his crops. In a unanimous decision the SCJN sided with the government, ruling that since the road linked two towns Terreros's concerns were secondary to the public utility of the proposed route. In 1936 the SCJN also ruled on Benjamin Bonilla's writ of *amparo*, which he filed after the federal government seized a road he owned in Mexico City. In a 3–5 decision, the judges found that a presidential decree to label the route a *camino vecinal* was legal.[16]

Nevertheless exceptions existed. In one case from 1939 the fight for open access to roads and land expropriation collided. Guada-lupe Ruíz brought a writ of *amparo* before the SCJN in which she said municipal officials in Tlaxcala violated her constitutional right to free movement when they closed a road that linked her property to the national highway from Veracruz to Mexico City. She called on the court to order state authorities to expropriate the land in question and repair the road, which had been torn up by farmers to plant maguey. Upon reviewing the case, the SCJN ruled unani-mously against Ruíz, concluding that the road in question could not be defined as a *camino vecinal*. Since the thoroughfare did not link two towns but only connected private lands to the national highway, the judges said it failed to meet the requirements for legal protec-tion. As a result the other farmers in the area were free to do what they wished, and the state was not compelled to expropriate any properties in defense of the route.[17]

Each of these cases, in different ways, reaffirmed state power over roads, giving the officials and legislatures greater latitude in defin-ing public utility, protecting open access to roads, and maintaining authority to expropriate lands. When private companies or landown-ers pushed back against this power, Cárdenas's SCJN consistently upheld lower-court rulings that defended the government. By doing so the federal judiciary protected ambitious state and regional road-building programs, which grew in size and complexity in the late 1930s as citizens demanded better quality motorways.

Changing the Structure of State Road-Building Bureaucracies

Thanks to strong federal backing through bonds, state agencies boosted spending on roads during the 1930s. Between 1933 and 1934 SCOP's cooperative fund for road construction more than dou-bled contributions, to 8.7 million pesos, with an annual growth rate that averaged 35 percent. In 1937, following the new round of road bonds Cárdenas offered, spending again spiked, rising 72 percent from the previous year to 25 million pesos.[18] Federal and state offi-cials in Nuevo León increased their financial commitment for the JLCNL by 50 percent to 1.4 million pesos as they launched a new six-year plan for road construction. This work included expanding and upgrading the motor route to the U.S.-Mexico border via Reynosa, Tamaulipas. Likewise in Veracruz the JCCV estimated a total agency budget of 2.46 million pesos, more than double what the state government had spent on road construction in the first half of the decade.[19]

Notable differences persisted in the ways the state governments of Nuevo León and Veracruz organized their road-building bureaucracies. In Veracruz the Junta Central de Caminos, DCOP, and the Junta Local de Caminos de Orizaba—organized by local business people—operated as separate entities with distinct budgets and funding sources. Each of these organizations also supervised different construction portfolios: the JCCV focused on projects designated as national highways, while DCOP worked on regionally oriented routes between small and midsize cities in the state. Meanwhile the JLC de Orizaba built and maintained the highway that connected Córdoba and Orizaba to the border with Puebla to the west and Veracruz port to the east.[20]

Nuevo León's bureaucracy remained much more centralized, relying on the JLCNL as the sole authority for all road building. In November 1935 Salvador Toscano retired from his position as technical chief of the agency, turning over duties to Pablo Domínguez Jr., an engineer who had risen through the ranks of the CCE and JLCNL. The new director maintained many of the day-to-day operational aspects of the organization Toscano had established, but also faced pressure from a newly formed road workers union to raise wages and improve benefits. In one key departure from Toscano's policies Domínguez allowed for greater participation from private contractors in regional road construction efforts, which reflected an overall shift in the balance of priorities for the JLCNL.[21]

Social and demographic factors fueled these distinctions. In Veracruz the state's much larger and less geographically concentrated population allowed for regional power blocs. The multipolar organizational structure of the state's road-building efforts reflected this political reality. In Nuevo León, Monterrey remained dominant in state politics. Moreover the JLCNL experienced much more consistency in top management; after Toscano's retirement, Domínguez helmed the agency for more than two decades. Conversely, there were notable amounts of infighting among Veracruz's road-building agencies, as well as high turnover in directors at the JCCV.

Nuevo León

Under Domínguez the JLCNL commanded an annual budget that averaged 800,000 pesos and employed thousands of workers. In 1936, amid growing complaints, personnel organized to demand changes to how the agency operated. In a letter to SCOP the new organization's secretary general, Manuel Saldaña, wrote, "We have been suffering from consistent delays in our wages, which oftentimes amount to three months of back pay owed to us. . . . Not only does this hurt workers, it hurts the reputation of the Junta Local de Caminos, adding to all of the miseries our families have experienced."[22]

The Sindicato de Empleados y Obreros Constructores de Caminos (SEOCC) became an important mechanism for workers to obtain more state support for themselves and their families.[23] The union endorsed a plan to issue new road bonds as a means to stabilize the JLC's payroll schedule. It also defended union members against unfair termination and fought to raise the minimum wage for *peones* and other staff.[24]

In 1937 Governor Anacleto Guerrero Guajardo cited many of the outstanding grievances raised by SEOCC when he pledged to extend to road workers all benefits afforded other industrial laborers via the 1931 Federal Labor Law. This promise included financial indemnities, free transportation for workers and families, seniority ladders, ten days of paid vacation for all employees, and a daily minimum wage of 1.5 pesos. The governor also announced significant changes to health coverage, guaranteeing a full-time professional staff of medical personnel attached to the JLCNL. A medical director and three assistants were appointed to make weekly visits to job sites. The agency began covering 75 percent of all medication costs for up to one year for technical and administrative staff, and 50 percent for laborers.[25]

These concessions did not defuse union demands, however. SEOCC leadership saw further negotiations as necessary to address longstanding grievances over low pay. In 1938 SEOCC argued that the governor's 1.5 peso wage was barely enough to cover living expenses. They demanded 2 pesos a day for *peones* and corresponding increases in wages for all other workers. Some within the government resisted the move, arguing that raising the minimum wage to 2 pesos would make JLCNL employees some of the highest paid road workers in the country. In January 1939, after further deliberation, the state government and SEOCC agreed on a compromise that raised the minimum wage to 1.75 pesos.[26]

One of the governor's policy changes that proved popular with rank-and-file employees was the creation of intramural basketball and baseball teams. JLCNL officials saw these groups as a means to improve morale, encourage good health, and reduce drinking.[27] By agreeing to these diverse concessions in favor of greater worker welfare, Guerrero had invariably reframed a major part of the JLCNL's mission. Alongside building roads, the agency became much more invested in providing workers with welfare services. When the JLCNL failed to fulfill expectations, employees voiced their disapproval vigorously through the SEOCC.[28]

As road workers won more rights and better pay inside the agency, Domínguez turned to the private sector to diversify construction efforts. He began evaluating private proposals for road building and coordinated with these groups in a supervisory capacity. In 1937 the Monterrey Iron and Steel Foundry and other local firms proposed construction of a new highway from the state capital to Mina, a *municipio* on the northwestern border with Coahuila.[29] In August, A. W. Villarreal framed the proposal as a public-private initiative, emphasizing broad support among local communities and transportation companies from across the region. He proposed the formation of a new committee to discuss logistical details and encouraged the JLCNL to begin surveying the route. Villareal also suggested charging a "subscription fee to local residents who could afford to pay" to help finance construction. He described a region ripe for industrialization and tourism: "I am confident that there are many people who want to help. . . . The road will reduce travel distances by at least forty percent. It will be extremely useful to Fábricas de Cemento [a private company that Villareal represented], the regional transportation companies, and the creation of future businesses that will certainly benefit from easier access to natural resources. It will also bring tourism to the banks of the Rio Salinas and to the nearby mountains with their mild climate and vistas."[30]

After a period of deliberation Domínguez agreed to go ahead with the project. No mention of whether the agency charged a toll for drivers is found in subsequent correspondence. Given the legal and political climate strongly in favor of open access to roads at this time, however, it is unlikely fees were imposed.[31]

In 1940 the JLCNL signed an important agreement with the Monterrey-based Compañía Construcciones Nacionales, S.A., that went beyond any previous public-private ventures. Whereas outside firms had operated as equipment and materials suppliers or contributed roadwork to existing state-run projects, the new contract put Construcciones Nacionales directly in charge of building additions to the Reynosa-Monterrey highway. Tomás Williams, the general manager of Construcciones Nacionales, already had experience working in Nuevo León's road-building sector. In 1935 Toscano had hired the firm's parent company, Carreteras de México, where Williams had previously worked, to build drainage systems to mitigate erosion along the Pan-American Highway.[32]

The structure of the plan was simple. Just as SCOP's DNC funded state road boards, the JLCNL apportioned funds to the company, which in turn used them to hire workers, buy equipment, and build the route. In July 1940 Construcciones Nacionales began with an initial budget of 40,858 pesos, which increased to an average 112,000 pesos a month for the rest of the year. The company modeled its personnel structure after the JLCNL. It divided operations into eight labor crews, employing more than 180 individuals, 78 percent of whom were *peones*. The remainder of the workforce included supervisors, machinists, drivers, and clerical staff in the field office. Most *peones* earned the 1.75 peso agency wage, while more experienced hands made 2 pesos a day. Crew supervisors received the highest per diem wage: 9 pesos. For equipment and material Construcciones Nacionales made all purchases through the JLCNL, which rented or supplied tractors, motor vehicles, tools, gasoline, and other items.[33]

Implementation of the road-building contract with Construcciones Nacionales revealed tensions between company officials and federal and state authorities. Following established accounting procedures, SCOP and the JLCNL required the company to provide monthly progress reports to justify expenses and make new financial requests. Soon after work began, however, Gilberto del Arenal, a top official at SCOP's Department of Cooperation, complained that reports for the Reynosa-Monterrey project failed to clearly itemize expenses: "To protect this board's interests, and the fifty percent contribution by the federal government, in my opinion, invariably, monthly statements must correspond with the contractor's activities."[34]

The department also wanted original copies of the company's incorporation papers and the contract it signed with the JLCNL for accreditation. During the fall of 1940 Domínguez was caught between an increasingly frustrated SCOP and a private contractor accused of ignoring repeated requests for information. In October, Próspero Castro, another high-ranking SCOP official, threatened to terminate federal contributions to the arrangement if nothing was done. Domínguez urgently wrote two letters, one to Castro, chiding the national bureaucracy for a lack of clear instructions on what information it actually wanted and for not having brought these concerns to him sooner. The second letter went to Construcciones Nacionales, ordering Williams to produce an immediate response: "I am writing for you to inform me—in the shortest time possible—about the status of this required documentation." Two days later, although making no apologies for the previous delay, the company finally sent the paperwork to the JLCNL for delivery to SCOP.[35]

These accounting and clerical problems indicate that the company may have been unprepared for the scope of planning and building such an important border highway. Earlier Domínguez had criticized the company for failing to conform to technical specifications agreed upon in the contract, noting that state engineers had found inconsistencies in the asphalt-concrete paving on the highway.[36] Notwithstanding these problems, neither the JLCNL nor SCOP pressed for the removal of Construcciones Nacionales from the project. This may have been due to the fact that the group's parent company, Carreteras de México, enjoyed close ties with national politicians, and JLCNL officials may have been loath to upset powerful friends of ex-president Rodríguez.[37]

Veracruz

Unlike in Nuevo León, the mix of agencies that carried out road-work in Veracruz fought over limited budgets and bureaucratic juris-diction. JCCV and DCOP officials in Xalapa complained internally about the slow progress of statewide construction efforts, arguing that the governor failed to provide sufficient funds. Mario Ojeda, the head of JCCV at the time, said the state was not doing enough to see the project to completion: "In 1934, no construction work was car-ried out due to the impossibility of the state government's finances, which have only covered the maintenance of what was built in 1933, and the gap [in the road] between Atzalán and Tlapacoyan." Oje-da's frustration summarized the political cleavages that remained a persistent challenge to road-building efforts throughout the 1930s in Veracruz. Engineers working different state-level projects often jockeyed for limited funds as routes deemed nationally important claimed the lion's share of financial support.[38]

By July 1936 the JCCV had spent more than 730,000 pesos on 81 kilometers of the Perote-Xalapa-Veracruz road. Despite this prog-ress, roadwork was already two years behind schedule. A key factor in causing this delay was a failure of the JCCV to adequately assess the impact of poor environmental conditions in formulating its con-struction timetable. Heavy seasonal rains in the state's mountain-ous terrain had washed out portions of the highway, which forced labor crews to backtrack and repair damage before moving forward.[39]

The JCCV's estimated unit cost for road building was 3.67 pesos, 32 percent more than the national rate of 2.5 pesos per cubic meter for construction of the Pan-American Highway. This contrast was due, in part, to the state's difficult terrain, but it may also have been a result of organizational problems. Personnel and management issues plagued the JCCV. In January 1937 the director, Miguel Cataño Mor-let, rebuffed attempts by DCOP to delegate more tasks to his agency, citing an already heavy work load. Later that year scandal rocked the

JCCV following two separate allegations of sexual assault by employees near work sites. Cataño Morlet ordered a restructuring of the agency's management to improve internal accountability. Then, in December, facing public criticism, the director abruptly resigned from his post after less than a year on the job. These problems took a toll on the agency, contributing to work delays and missed deadlines in its construction schedule.[40]

In contrast to the JCCV, officials with the JLC de Orizaba appeared to run a much more efficient and cost-effective operation. The organization had been set up by local business leaders who wanted to expand road access to Orizaba and its sister city, Córdoba, in central Veracruz.[41] In 1936, as part of Veracruz's six-year plan for road construction, the agency supervised work on the 117-kilometer Puente Nacional–Huatusco-Córdoba Highway, which joined the national road en route to Veracruz from the southwest. Based on monthly expenditures, this asphalt-concrete route carried an estimated unit cost of 2.33 pesos per cubic meter, slightly below the national average.[42]

Miguel Alemán's arrival as governor of Veracruz at the end of 1936 led to new policies that improved financial conditions. He increased the statewide road-building budget to 2.1 million pesos through the issue of new bonds and also tolerated the creation of more local road boards to take the lead in carrying out construction efforts. For example, in 1938 a new committee was formed in Coatzacoalcos charged with construction of a coastal highway to Minatitlán. Following the model pioneered in Orizaba, it worked independently from the JCCV and DCOP. The group set up its own management and personnel structure, financing operations that averaged 20,000 pesos a month via payments from the regional business community. It employed, on average, ninety workers to staff construction crews. The new organization also paid personnel differently than other state agencies; *peones* received 2 pesos a day, which was 25 cents more than the JCCV's standard wage. While the Coatzacoalcos-Minatitlán committee sourced much of its financing, labor, and management locally, the state did supply it with a supervising engineer who handled day-to-day technical matters.[43]

In rural, isolated parts of the state, *ejidal* (communal land) committees often took charge of construction and maintenance efforts. For example, in September 1938 more than a dozen groups with membership in the League of Agrarian Communities and the Farmers Unions from Zongolica, in southern Veracruz, convened a special meeting to build a framework of demands to deliver to DCOP. Constituent committees insisted on more equipment from the state to open and repair roads. They also wanted a guarantee from the governor that public authorities followed through on any promises made. In October, Anastasio González, president of the Rancho Nuevo *ejidal* delegation, requested that DCOP provide hand tools for road repairs. The department forwarded the query to Adolfo Omaña, director of the state office for indigenous affairs, who supplied twenty used pickaxes and shovels. On November 28 Omaña wrote a local official in Zongolica of the decision: "This equipment . . . will be used exclusively for highway repairs, so I recommend you safeguard its intended use, but of course, it will ultimately be the farmers' responsibility."[44]

Cost savings are difficult to calculate due to a lack of financial data included in the official correspondence with *ejidatarios* (communal farmers). This case, and related ones, suggest that the state had reduced expenses for road maintenance by supplying used equipment to locals. There is no indication that many *ejidatarios* were ever compensated beyond the tools that officials contributed. As such, although the state had largely ended official use of the *faena* to build highways, aspects of it may have remained as an unstated policy for local maintenance efforts well into the 1940s.

At this time no formal labor organization that represented road workers existed in Veracruz. Instead employees used an existing network of agrarian and labor associations to petition the government over demands. Even if a road workers' union existed, given the lack of a unified management structure for road building, it would have been difficult to coordinate pay scales and labor policies across all of the agencies.[45] Moreover Veracruz already enjoyed a strong and established base of popular political activism. For years agrarian activist groups had organized *peones* who worked in Veracruz's road-building agencies. Meanwhile, in Orizaba the local chapter of a state union associated with the influential Confederación Regional Obrera Mexicana (Regional Confederation of Mexican Labor, CROM) coordinated with the municipal road board to ensure that enough workers were available for construction projects along the route for a road being built to Huatusco.[46]

CROM and the League of Agrarian Communities acted as political and negotiating partners for a large number of smaller local organizations. In 1937 *La Voz del Campesino*, a radical newspaper published in Xalapa, counted nearly three thousand different *ejidal* commissions and popular organizing committees, including twenty-seven regional unions for workers and campesinos across the state. In addition to these groups, transportation cooperatives also represented road workers in disputes with the state government over wages.[47]

This kind of political engagement existed, in large part, thanks to the legacy of Tejeda's support for agrarian activism. Existing rural and labor organizations had accumulated considerable influence and played an integral role in state road-building efforts over decades. Throughout the 1930s the state continued to work closely with CROM, appointing its members as labor representatives on road-building boards. Nevertheless the groups that had allied themselves with Tejeda gradually lost power. Succeeding governors turned to Vicente Lombardo's Confederación de Trabajadores de México (CTM) to form new labor partnerships. By the early 1940s, when the country entered the Second World War, the CTM became one of the principal federal labor organizations, operating through regional unions such as the Federación de Trabajadores del Estado de Veracruz, to represent road workers' needs in the state.[48]

Popular Demand for Access to Motor Travel

Between 1934 and 1940 an estimated 79,000 automobiles circulated nationally. Most people, however, continued to rely on bus companies and rural transport cooperatives to access the country's burgeoning road network. Holiday travel became one important source of income for many of these organizations as working-class and middle-class tourists from Mexico City took advantage of vacation deals and extended weekends to visit nearby destinations. For example, during the Easter holiday of 1934 a record numbers of cars and buses took to the highways bound for Puebla, Cuernavaca, and Acapulco. Companies sold nearly 14,000 tickets for the weekend, running 250 trips daily to these cities. To meet high demand, managers reached out to other city businesses, renting any available trucks to carry so many passengers.[49]

By the end of the decade new firms established more regional and long-distance bus routes and reduced fare prices. In 1939 Autobuses de Oriente opened for business, launching the first full-length route from the port of Veracruz to Mexico City. It also offered second-class, round-trip fares between Xalapa and Veracruz for 3.6 pesos. Transportes Frontera started running three daily trips from Laredo to Mexico City, with a stopover in Monterrey. It charged 26 pesos for one-way tickets, notably lower than competing fares from previous years.[50]

The growth strategies that some bus companies implemented quickly saw success in the travel market. In 1937 in Nuevo León the Sociedad Cooperativa de Transportes Monterrey-Cadereyta-Reynosa, which had begun with seven vehicles, soon doubled its fleet to fourteen, offering ten daily round-trip routes from the border to Monterrey. That same year the radical-minded founders of the Unión de Camiones de Orizaba y Anexas paid off the loans they had incurred to purchase the cooperative's vehicles. Employing up to seventy drivers and mechanics, the group outgrew its original headquarters and had begun looking for a new location to further expand the business.[51]

Rural communities relied heavily on transport cooperatives to meet their motor travel needs. Besides carrying passengers and goods, these services took on political roles, advocating for roadwork. They actively petitioned officials to build more roads and fund maintenance efforts. For example, in Huatusco, Veracruz, the local transport union wrote President Cárdenas, requesting emergency funds to finish a road to the town after the local crew had run out of money. They argued that without this help, residents would have to travel the remaining 25 kilometers on horseback to deliver goods to the train depot. The president's office agreed to the request, and work restarted soon after.[52]

This example speaks to a larger trend wherein bus companies helped to expand intermodal transportation options for rural citizens. They connected remote communities to distant railroad networks that had previously been reachable only via animal conveyance. In doing so they also added to calls to continue developing the nation's system of highways and state roads, reduce travel times, and provide low-cost fares. By the end of the 1930s Mexico had experienced a 40 percent increase in bus travel with more than 10,000 vehicles in circulation nationwide, serving urban and rural destinations.[53]

Oil Expropriation and Its Impact

The gradual economic recovery under way during the middle years of Cárdenas's *sexenio* drove political optimism in support of developing motorway infrastructure. The president had already made clear in national addresses that building new roads provided capacity for greater levels of domestic and foreign motor tourism. At the same time, however, political and social tensions were growing over the nation's economic future and its ability to control natural resources, especially oil.[54] In May 1937 17,000 oil workers launched a twelve-day strike that stopped gasoline production and shut down motor transportation across the country. Gas stations and refineries closed as strikers took to the streets carrying red-and-black flags and demanded a 6-peso daily wage. In Mexico City half of all buses were out of service, and automobiles with empty tanks were abandoned along streets. Hundreds of U.S. tourists were stranded, urgently trying to contact U.S. authorities for help to return to the northern border. By the second day of protests the Mexican Automobile Association had acquired emergency gasoline shipments from the United States, which it provided to desperate motorists. In Laredo border control stations recorded only fifteen motor tourists crossing daily into Mexico since the strike started, down from an average of five hundred.[55]

Although public opinion largely disapproved of the industry walkout, this event foreshadowed looming political battles with the foreign oil companies that would marshal popular support for the strikers. On 18 March 1938, after U.S. and British managers refused to respect a SCJN ruling, Cárdenas announced that he was taking steps to nationalize the petroleum industry. In a public address following the decision he stated, "The oil companies, despite the government's best attitude and considerations, have waged a deft and skillful campaign . . . [to] inflict serious economic injury to the nation and to nullify the legal determinations made by the Mexican authorities." Thousands of citizens took to the streets in support of Cárdenas's decision as foreign governments called for a general embargo on Mexican products. Even the Pan-American Highway became a space where popular sentiments manifested against the U.S. economic sanctions that followed. In August billboards appeared on the border road declaring, "The money you spend outside of Mexico, your children will not have later."[56]

The expropriation issue, in conjunction with shortages brought on by the U.S. oil embargo, affected motor travel and federal road construction efforts. Between 1935 and 1937 the size of the national road network increased by 43 percent, from 4,260 kilometers of highways to 7,510 kilometers; in contrast, from 1938 to the end of 1940, this rate of growth fell by a quarter. Likewise foreign tourism decreased by 20 percent, while transportation companies faced higher fuel and equipment costs due to a weak peso.[57] Across the country, including in Monterrey, bus companies fought with municipal officials over price caps on fares.[58]

Despite these problems the embargo appeared to have only a limited effect on state-level road-building activities. Amid the brewing oil crisis the JLCNL launched work on the Monterrey-Reynosa Highway, employing Construcciones Nacionales as a principal contractor, while the road workers' union won a wage hike. In Veracruz regional construction efforts on the national highway continued apace, showing increased spending during Alemán's tenure, in which work on the routes to Veracruz port and Tuxpan was finished. Furthermore, although no federal road bonds were issued in 1938, Cárdenas announced new rounds of offerings in 1939 and 1940. Under his successor this financing program became an important area for bilateral negotiations between Mexico and the United States as the two countries repaired their strained economic and diplomatic ties.[59]

The Cárdenas years represented an important period of transition and growth in the national and state bureaucracies that built roads and promoted motor travel. The president expanded policies his predecessors had established, while issuing bonds that generated millions of pesos for construction efforts. He also appointed an ideologically powerful Supreme Court that challenged private property rights in favor of greater road building and public access to roads.

Some of the most notable changes that occurred during this period came at the state level. Nuevo León's authorities reworked the objectives of its road-building agency, fashioning it into a paternalistic organization with a wider mission that included social programs for workers. In Veracruz, although state agencies like the JCCV continued to struggle to carry out its activities, the government allowed local communities to launch regional initiatives to build highways and *caminos vecinales*. The following chapter examines how these organizations navigated problems of even greater economic scarcity as Mexico entered the Second World War, while state governments faced a more assertive SCOP under the leadership of the venal and garrulous Maximino Ávila Camacho.

4

"We March with Mexico for Liberty!"
Road Building in Wartime

In December 1941 Veracruz's governor Jorge Cerdán delivered the state road-building budget to President Manuel Ávila Camacho. The plan earmarked 2 million pesos for construction efforts the following year, including a major expansion of the Minatitlán-Coatzacoalcos coastal highway to Tampico, Tamaulipas. Cerdán wrote of the road's "strategic value to the White House's plan for continental defense," adding that it could be used to quickly mobilize troops against a potential Axis invasion. He also emphasized Minatitlán's and Tampico's military importance as cities with major refineries and other facilities that produced high-grade lubricants for aviation and heavy machinery. The 400-kilometer highway would serve as the backbone for these operations, carrying much-needed supplies from across the region.[1]

World War II was a prodigious time for road building, marking important changes in Mexico's domestic and foreign politics. The new government under President Ávila Camacho (1940–46) overcame acute economic hardships, advancing construction efforts with a combination of deficit spending, foreign-direct investment, taxes on gasoline, and (often haphazard) rationing of basic commodities. It repaired diplomatic relations with the United States, which opened up millions of dollars from Washington in investment for road building. From 1941 to 1946 the national highway network grew by more than 33 percent, as the country added 1,250 kilometers and spent

50 million pesos annually. The government also increased spending for road maintenance, allocating nearly 100 million pesos for more than 26,000 kilometers of repairs.[2]

Waste and graft, however, presented serious complications to the national road-building program. Despite the federal government's impressive figures for construction and maintenance efforts, rural communities complained about a lack of official support for upkeep programs. They identified roads that became impassable during bad weather and others that had gone years without a visit from maintenance crews. Citizens also criticized state authorities who were slow to address problems that made roads dangerous for the people who lived along them. These types of concerns called into question where the money allocated for roads may have actually gone.[3]

In a speech before the Chamber of Deputies, Ávila Camacho declared, "It is not possible to truly integrate a sense of the nation without an ample road network that facilitates economic exchange [and] connects people."[4] By equating road building with nation building, he reaffirmed that improved transport mobility was essential to economic growth. His government defended regional bus services from state-level meddling, while road-building agencies played a key role as mediators in local political fights over land use. Nevertheless the president's rhetoric may have rung hollow for some citizens when he allowed his notoriously corrupt brother, Maximino, to take control of SCOP and appointed judges to the Supreme Court who favored private interests in disputes over public claims for new motorways.

Ávila Camacho used the money that financed road-building efforts as a key part of his plan for national reconciliation. Following the contentious 1940 election cycle, he sought to resolve tensions between the business community, government, and workers. He favored political moderation over the perceived excesses of Cárdenas's socialist policies, appointing Supreme Court judges who strongly supported private property rights in cases that involved land expropriation for road building. In other ways the new political direction was felt at the regional level as state governments centralized decision making in the road-building bureaucracy. They passed new rules outlining

public use of road spaces and carefully defined how regional organizations requested funds and participated in construction efforts.
In Veracruz state officials further empowered local road-building
committees to have a say in planning and construction efforts, but
this freedom came with renewed scrutiny and close supervision.
As states asserted control over the mechanisms of road building,
local actors turned to the courts and sympathetic politicians and
even exploited some government rules to their advantage to fight
against official overreach.

These activities played out against a wider backdrop of national
wartime mobilization. Mexico's entry into World War II on the side
of the Allies affected road-building efforts in distinct ways. Public
officials faced tightening state budgets and material shortages, while
ordinary citizens suffered austerity measures and soaring prices that
limited options for motor travel. In response to these everyday challenges amid the impact of global conflict, Ávila Camacho's political, legal, and economic reforms irrevocably changed how Mexico
funded, built, and used new roads. By the end of the war the country had shifted firmly away from the populist, left-wing politics that
had marked the Cárdenas years and had moved toward more conservative, pro-business policies that gave less direct support to workers and farmers.

Mexico's Road to War

The Roosevelt administration urgently wanted to resolve the petroleum issue to ensure access to Mexico's natural resources in the
event that the United States entered World War II. In February 1941
it forced Standard Oil and the rest of the industry to accept a one-
time payment of 37 million dollars from the Mexican government
for their holdings lost in the 1938 nationalization. This deal marked
the start of a series of negotiations on trade and industrial development, culminating in the signing of a new bilateral economic accord
in November. The agreement provided significant U.S. investment
in the Mexican economy via the Export-Import Bank, including
a 10-million-dollar bond for road building. Later the Ex-Im Bank
added another 20 million dollars in loans for new motorways. This

money guaranteed work continued on the Pan-American Highway to Mexico's southern border with Guatemala and also provided for the purchase of new heavy equipment for other road construction and maintenance projects.[5]

For its part, the Cárdenas administration had already cooled to the idea of seeking out alternative economic partners. In August 1939, although it had signed a deal to barter oil for industrial equipment from Nazi Germany, Mexico halted any future negotiations on oil deals, citing the deteriorating political situation in Europe. Then, in October 1940, it canceled an industrial agreement with Tokyo, which, among other provisions, had brought a small group of Japanese road surveyors to Mexico to work with SCOP. President Cárdenas described the decision to cut ties with the nations of the Anti-Comintern Pact as "an act of continental solidarity" with the United States and its allies. Moreover the change in tone that Cárdenas initiated and Ávila Camacho continued signaled to U.S. Americans and conservative elites in Mexico that national leaders did not want to abandon commercial ties with Washington.[6]

In May 1942 Mexico declared war on Germany and the other Axis powers, following the sinking of two PEMEX tankers, the *Potrero del Llano* and the *Faja de Oro*, by a German submarine. Political leaders rallied the nation to support the Allied cause. In June, after Mexico recovered the remains of the sailors from the U.S. Coast Guard, the government launched a highly publicized caravan tour to deliver the bodies from the northern border to Mexico City. Trucks carrying the caskets passed through numerous towns and cities on the way, giving ordinary citizens an opportunity to feel connected with these larger events. By doing so public officials transformed the Pan-American Highway into a kind of public stage that framed the official wartime narrative, encouraging citizens to express anger over the attacks. People in rural communities lined up along the highway to catch a glimpse of the vehicles as they drove by. Public rituals marked the caravan's passage; in Monterrey large civilian gatherings filled the city's streets and parks as the trucks arrived for a special ceremony at the cathedral.[7]

Wartime political propaganda urged national cohesion and social and economic mobilization. It evinced a general rightward shift in tone and policy that would go on to affect much of society. *Excelsior* depicted Ávila Camacho as a paternal guardian of a nation at the vanguard of Latin American countries united with the Allied forces against the Axis powers. The weekly magazine *El Tiempo* described U.S. investment in road building and industrialization as a means to transform Mexico's countryside and improve the living standards of average citizens. State governors printed obsequious statements of support for the president in national and regional newspapers. Governor Cerdán declared that all of Veracruz was behind Ávila Camacho, calling for "noble and humanitarian political unity during this time of responsibility."[8] Likewise the *Gaceta Oficial de Veracruz* declared, "We are at war! Farmer: sharpen your sickle. Worker: take up your hammer. . . . Industrialist: increase your production . . . and in this way, united in supreme effort, wills strained, firm in purpose, we march with Mexico for liberty!"[9]

Industries associated with motor travel reinforced calls for greater social and economic sacrifice in the name of patriotism. With the slogan "Take care of your car or walk," PEMEX touted the importance of good lubricants to ensure that engines lasted until war's end since spare parts would be very difficult to obtain: "Such care is unnecessary under normal conditions, nevertheless this current emergency not only justifies, but demands it."[10] An advertisement for Veedol Lubricants depicted soldiers operating a howitzer cannon in the heat of battle with the headline "Here's the 1943-model car you didn't buy!" The caption reminded consumers that sacrifice was "a vital part of the struggle towards victory. You have the right to feel proud that you're contributing to this cause." By conflating the decision to forgo an automobile purchase with the need to reserve materials for the production of heavy artillery, these kinds of narratives tied automobility to wartime scarcity. Denying oneself a car became symbolic of a patriotic and self-sacrificing commitment to the national cause.[11]

The political shift in emphasis to military preparedness against the Axis powers had significant repercussions on civilian motor travel. Tourism suffered as travel restrictions reduced cross-border traffic, forcing hotels, restaurants, and other businesses on the Pan-American and Inter-Oceanic highways to close or reduce operations.[12] Mexican and U.S. officials increasingly viewed the borderlands as vulnerable to infiltration or invasion by the Axis powers. They coordinated on counterintelligence measures to restrict the movement of German and Japanese citizens in the region. The U.S. Army also proposed the construction of new motor roads and other transport infrastructure to facilitate troop deployment in the event of an attack, but Mexican authorities were reticent to approve the idea due to sovereignty concerns.[13]

Road engineers faced wartime gasoline shortages that threatened to delay construction timetables. In Nuevo León JLCNL officials estimated that the agency needed 20,000 gallons of petroleum to sustain operations, but federal rules prioritized military needs over all others. Official guidelines for state agencies on how to requisition basic commodities like gasoline were murky at best. As a result coordination efforts between federal and state authorities on logistical issues, which had long worked smoothly, became strained. As there was not enough supply to meet official demand, JLCNL engineers competed with urban transit authorities and other JLCs for limited materials. In 1944, fearing they would be forced to suspend operations due to shortages, agency officials reached out to U.S. energy brokers in Brownsville to secure a contract for sufficient oil reserves.[14]

Shortfalls in tax revenue quickly led to other fiscal challenges. In August 1944 the JLCNL wrote to the Dirección Nacional de Caminos to request an additional 50,000 pesos monthly from SCOP's cooperative fund to cover anticipated budget deficits. When that appeal proved unsuccessful, the governor intervened, writing to SCOP about gaps in the JLCNL budget due to wartime economic disruptions. The DNC's Department of Cooperation rejected the governor's entreaty, stating that all available monies for the upcoming year were already allocated, and nothing more could be done.[15]

The administrative tensions that emerged over wartime scarcity affected how public officials prioritized funding of roads, citing usage concerns and a mistrust of counterparts in other states. In May 1944 Governor Arturo de la Garza delayed approval of Nuevo

León's budget contribution for a new road to the state border with Tamaulipas, wanting to ensure that the neighboring state had also allocated enough money to continue construction. If not, he would instruct the JLCNL to immediately stop roadwork at the town of General Treviño, 17 kilometers from the state border. His advisors had warned that the road would be underused if Tamaulipas did not follow through with construction. In response SCOP reassured de la Garza that Tamaulipas would honor its part of the plan, and roadwork ultimately continued.[16]

Shortages disturbed everyday mobility across the nation and also encouraged corruption. By the end of 1942 spare parts for automobiles, buses, and trucks were scare, and market speculators drove up the cost of many items. Despite claims by carmakers that prices for parts had only risen between 6 and 10 percent, the problem was much worse. For example, drive chains and sprockets that previously cost 132 pesos now sold openly for 350 pesos; likewise gear pumps and rolling bearings rose from 37 and 11.5 pesos to 49 and 24 pesos, respectively. Tire rationing became particularly acute. In 1943 a single tire cost 550 pesos on the black market, almost double the price before 1941, representing more than a month's wages for a midlevel professional worker.[17]

These problems forced bus companies and transport associations to dramatically reduce service. The Confederación Nacional de Cooperativas (National Confederation of Bus Cooperatives), which included 276 bus transport groups, estimated that its members had retired 2,266 of more than 6,800 vehicles nationwide due to the severity of shortages. In late 1942 bus companies submitted monthly requests for replacement tires to the government that totaled 4,400 units nationwide, but public officials could supply only 600 units to meet this demand. By the spring of 1943 more than 240 of the 785 passenger buses that operated in Veracruz, Puebla, and Tlaxcala had been sidelined, while in Nuevo León firms removed 190 of 300 motorbuses from daily circulation due to shortages. Some cooperatives turned to the black market, relying on anonymous sellers to buy tires, gasoline, and much-needed repair parts.[18]

Bus companies and cooperatives pressed public officials to allow fare increases. In Monterrey they asked to raise ticket prices from 5 to 10 cents, citing the pressures of high fuel and maintenance costs on their businesses. When municipal authorities refused, some groups hiked fares anyway, only to be fined and have their drivers detained and vehicles impounded. The Alianza de Camioneros de México, a national trade group, sued municipal and state governments on behalf of these bus operators, arguing the 5-cent fare was pushing its members into bankruptcy. The case went before the Supreme Court, which ultimately ruled against the fare hike, concluding it was unfair to local citizens in wartime.[19]

In 1943, in an attempt to address the shortages afflicting the transportation sector, national authorities declared rubber a protected commodity. Although this decision helped to boost domestic tire production, problems persisted. In August 1944 a dispute with tire makers over pricing nearly erupted into a general strike when the National Confederation of Bus Cooperatives threatened to halt service across the country. President Ávila Camacho urged both sides to show patriotic goodwill as the Secretaria de Economía Nacional hastily convened negotiations to address the issue. The agreement to form a special commission to respond to the confederation's grievances narrowly averted the work stoppage.[20]

Many local communities also pushed state governments for road-building resources during the war, using cultural and economic themes to articulate their needs. Rural pro-road groups emphasized the logistical challenges to motor travel, which they asserted had hurt regional economic development. For example, in July 1943 the mayor of Huiloapan, a hamlet of fewer than five hundred residents in central Veracruz, discussed the community's history and saw road building as necessary to connect to outside markets to attract visitors. In a letter to Rubén Bauchez, DCOP's tourism chief, he wrote:

> Perhaps you know that this pueblo is one of the oldest in the state . . . originally called "Ostotipec." [It is where] Hernán Cortés, during the expedition to Honduras, stopped to marry La Malinche to Captain Juan de Jaramillo. The marriage . . . occurred in the chapel here, and from that time forward the town has been named "Huilo-apan" which

means: "Doves together in the water." Over the years, however, this place has been forgotten and has not progressed . . . because enough has not been done to publicize it. Mr. Secretary, we await your help for the good of this community; with national and foreign tourism, all manner of cars could pass through this area if it were not for the terrible conditions of the highway.[21]

Whereas national and state authorities had altered their political rhetoric to emphasize wartime themes, many rural leaders continued to center appeals in the language of basic needs and tourism. They highlighted reduced isolation, the ability to transport goods more easily, and the potential for sightseeing trips as chief benefits of road building.[22]

In April 1943 in El Ojital, Veracruz, the Comité Municipal Antinazifascista wrote Governor Cerdán about disruptions the war had caused. They wanted him to intervene with PEMEX, which had delayed delivery of asphalt and concrete to pave the local road to Tuxpan. "Our ability to conduct the harvest is being gravely damaged," explained the committee's president, "due to the lack of an easily-traversable route."[23] Similarly in March 1944 in Papantla, Veracruz, the local chamber of commerce wrote the governor about building an extension of the Mexico-Tuxpan Highway to their town. The group argued it would allow residents "to conveniently take advantage of all of the commercial, agricultural, and tourism opportunities that this highway has to offer."[24] While local and national actors espoused different end goals in support of road building during the war, their agendas shared elements that allowed for some cooperation.

Reforming Federal Road-Building Policy

In October 1941 the president's older brother, Maximino Ávila Camacho, became the new head of SCOP. In a hard-nosed political takeover of the organization, the elder Ávila Camacho was said to have walked into the SCOP secretary's office and taken it "on a good day with his pistols," removing his predecessor, Jesús de la Garza, without fully consulting his brother. In Puebla, where Maximino had served as governor, he was known for this kind of brash approach, using aggressive and violent tactics against rivals. One of the first decisions he made as chief of SCOP was to replace all active federal directors at the agency, installing a new generation of assistants loyal to him. These arrivals included Carlos I. Betancourt as deputy secretary, a member of the elder Ávila Camacho's inner circle in Puebla.[25]

Maximino took a tough stance against state requests for more road-building funds. During his tenure SCOP repeatedly delayed the federal portion of wage payments to JLCs, despite heavy U.S. financial support for wartime Mexican infrastructure development. It was a largely unprecedented move, which may have been a tactic used in retaliation against governors who angered the elder Ávila Camacho for making what he perceived to be too many demands on SCOP resources.[26]

In his initial weeks at the agency, Maximino carried out a national inspection of the country's highway network. He met with state officials to discuss budgets and logistical needs for new roadwork, likely signaling to regional elites his desire to take a more active approach in running SCOP. His first stop was Monterrey. In a speech to local leaders, he reiterated national unity, pledging that SCOP would be a partner "irrespective of party affiliation."[27] The visit marked an important first step in President Ávila Camacho's offensive to promote political conciliation following the divisive 1940 election. It also signaled that the Ávila Camacho brothers were firmly in charge of SCOP and its national programs. Monterrey had been a stronghold of support for the opposition candidate, Juan Andreu Almazán. In an article published just before Maximino's arrival, *El Norte* touted his qualifications, challenging accusations of nepotism and, of course, making no mention of the man's popular reputation as an obstreperous womanizer:

> As we all know, the public always interprets the government's actions in the worst ways. The appointment of the president's brother to carry out such an important post at the Ministry of Communications can be seen purely as nepotism, but that is not the case here. On the contrary . . . everyone remembers that this new minister—during

his tenure as the governor of Puebla—was applauded by all of society for what he achieved . . . bringing wealth and prosperity to the state . . . attacking the false promises of the old political bosses . . . and forcing them to follow the law. A man with these qualifications deserves to be named minister in any administration that wants to take advantage of men who have worked for the good of the nation.[28]

This attack on the "old political bosses" took different forms and indicated the new administration's desire to reach out to powerful groups that had been typically hostile to the federal government under Cárdenas. The nationalist policies and patriotic rhetoric President Ávila Camacho marshaled with Mexico's entry into World War II reflected deeper right wing political and legal changes under way.[29]

The new administration not only asserted its control over SCOP but also made notable changes to the ideological composition of the Supreme Court. The judges that Ávila Camacho selected presided over an important shift in jurisprudence on road building. Whereas the court under Cárdenas had consistently ruled on behalf of state interests over private property rights, this changed in the 1940s. Reflecting the new president's more conservative policy of national conciliation, his appointees to the Suprema Corte de la Justicia de la Nación (SCJN) sided more often with landowners and businesses to the detriment of regional road-building schemes. Of five cases that came before the court that addressed public utility and land expropriation for road construction, the judges sided with the state only once. Most notably the SCJN restricted how state officials could seize land for road building.[30]

By limiting the power of state governments, Ávila Camacho used the courts to bolster his administration's agenda, mollifying conservatives. He could afford to deny governors the free hand that Cárdenas had allowed them in road-building cases because they relied heavily on the national coffers for funding and materials in a wartime environment that stressed political unity. The chief benefit was that landowners and private companies, typically on the losing side of SCJN decisions in the 1930s, saw Ávila Camacho as someone who listened to them. He could point to the ideological shake-up in the court as a tangible example of his willingness to address the concerns of conservatives and businessmen.[31]

The new SCJN appointees set a high bar for state government claims on the necessity of land expropriation for road-building projects. For example, in 1942 Gabriel Domínguez brought a writ of *amparo* against SCOP and the DNC over plans to build a new road near his property in the state of Guerrero. The agencies defended their work, arguing that since the road served the public good, building it superseded any individual objections. In a unanimous decision the SCJN rejected this assertion, ruling that the government had failed to actually show the public utility of the proposed route. The court upheld Domínguez's writ of *amparo,* and in doing so challenged a decade of legal decisions that had favored federal and state governments. The SCJN later reaffirmed its position that the authorities had to show greater evidence of public utility in order to expropriate land for road building. In 1944 the judges unanimously upheld María Garaby Cuevas's writ of *amparo* in Michoacán, concluding that state and municipal officials could not seize portions of her property since they failed to prove the public utility of a proposed *camino vecinal.*[32]

The Supreme Court found other ways to limit legal powers agencies had heretofore exercised over road building and public access to motorways. In 1942 the second chamber upheld the writ of *amparo* for an auto-transport cooperative that served central and northern Mexico. State authorities in Morelos had sought to restrict the group's ability to operate locally, ordering them to reduce bus service in the *municipios* of Jonacatepec and Axochiapan. The SCJN ruled that since the cooperative's buses used federal highways and state *caminos vecinales,* they were constitutionally protected, allowing for the free flow of traffic between towns and cities. In a separate case in 1941 the court also unanimously agreed that individuals could bring forward writs of *amparo* on behalf of communities against attempts to restrict regional mobility on roads, even when state or municipal officials declined to do so.[33]

Even as it restricted state government power, the SCJN reiterated the federal government's ability to police national highways. In Puebla and Veracruz, Gregorio Ortega Brito and other plaintiffs disputed whether national authorities could fine bus companies that used the Mexico City–Veracruz Highway. They claimed that service infractions SCOP levied against them violated their constitutional right to freedom of movement. The group's lawyers said Ortega and his associates had operated their bus routes in the same way since 1932 without any previous incidents with the authorities. Nevertheless the SCJN ruled against Ortega and the others, saying that SCOP and the Federal Highway Policy had the jurisdiction to enforce transit rules and impose fines for service violations in the interests of public safety.[34]

These verdicts established important legal distinctions between federal and state power. The SCJN interpreted the national Constitution in a way that limited state and local officials from controlling motor travel but did not fully extend this restriction to federal authorities. In doing so Ávila Camacho's Supreme Court permitted private citizens and businesses to defend their lands and operations so long as they did not infringe on federal supervisory powers like policing. This arrangement created a modicum of legal space for plaintiffs to win cases against state and local officials.

State and Local Road-Building Efforts under Ávila Camacho

The administrative and legal reforms that affected SCOP and the SCJN spoke to the political consolidation of these federal institutions by Ávila Camacho and his cabinet. They reflected the president's broad strategy of national conciliation and the important role that roadbuilding policy played in this process. However, the new administration went only so far in implementing these changes with the states. Despite Maximino's quick shake-up of SCOP, the federal government did not force similar modifications on state-level bureaucracies. Rather than risk further alienating powerful regional political allies, the president did not interfere with how state governors carried out planning and construction programs. This approach allowed for continued experimentation in how state engineering bureaucracies functioned, where notable variations persisted in how officials carried out their work. Nevertheless, tensions emerged over federal contributions to state spending on roads.

Nuevo León

Between 1940 and 1946 the JLCNL made few major changes to its hiring and labor practices. Employment rosters indicate that field crews remained between thirty to fifty men, most of whom were engaged in clearing and paving routes, while one to three engineers supervised construction brigades made up of at least half a dozen teams. Monthly salaries paid to top staff and engineers ranged from 350 to 750 pesos, while *peón* wages rose slightly, to 2 pesos per day, by the mid-1940s. The JLCNL also continued its program of paternalistic social activities intended to foster esprit de corps within the agency. Chief Domínguez's personal correspondence showed that he sent out wedding and baptismal congratulations to workers and their families, wrote letters of recommendation for specialists and *peones* alike, and continued to authorize the popular intramural baseball team, Los Caminos del Estado.[35]

The activities of the Sindicato de Empleados y Obreros Constructores de Caminos (SEOCC) likely ensured that the JLCNL remained responsive to personnel needs. The union continued to advocate for expansive medical assistance for grievously injured employees beyond the agency's standard coverage. For example, in August 1942 Francisco Herrera, a crew supervisor with a distinguished fourteen-year record at the JLCNL, suffered a work-related accident on the road from Monterrey to El Jabalí when he lost control of the truck he was driving and the vehicle crashed. First responders stabilized his condition and transferred him to a hospital in the state capital, where doctors amputated part of his left leg. While Herrera convalesced, the union wrote to the JLCNL as well as Nuevo León's governor Bonifacio Salinas (1939–43) urging for more to be done for the man: "Our comrade finds himself in difficult financial conditions due to the needs of his wife and eight children. We ask of you . . . in a humanitarian sense that you provide him with a just amount of assistance, which is his right by federal labor laws."[36]

Domínguez agreed. The agency moved Herrera to Mexico City for months of additional therapy, including fitting him with an artificial limb; it also provided financial help to the family. The following year the director received a letter from Herrera personally thanking him for the help: "I'm already getting along in my recovery and . . . the great part of all of this I owe to you for the concern you took in my health. For that I give you my sincere thanks as well as my children's appreciation for all that has been done for me." Herrera's expression of gratitude to Domínguez, viewed in the context of union activity, highlights the delicate balance that workers and agency managers tried to forge. Operating within the paternalistic framework of the JLCNL, the union pressed the institution for greater welfare concessions, all the while doing so in a manner that credited the state bureaucracy.[37]

In 1940 and 1941 Nuevo León was spending more on the JLCNL than ever, which may have afforded Domínguez the power to extend additional worker benefits when deemed necessary. Upon taking office, Governor Salinas made road building a top priority, growing state contributions to the agency's budget from 900,000 pesos in 1939 to 1.2 and 1.5 million pesos in 1940 and 1941, respectively. He hoped to use a 10-million-peso road bond, initially approved by the state legislature in 1941, to support further spending but needed to gain the approval of the Secretariat of Finance to do so. In March 1941 he dispatched Prisciliano Elizondo, an agent of the Fundidora de Fierro y Acero de Monterrey, S.A., to locate a bank in Mexico City to act as fiduciary for the state bond. This venture indicated the willingness of public and private groups in Nuevo León to work together to achieve shared objectives. Soon after Elizondo arrived in the national capital, he won an agreement to back the bond.[38]

Although Salinas had the support of commercial backers in the state, he faced tension with the national government that had a real impact on how the JLCNL operated. By the spring of 1942 Salinas had again increased the state budget for new roads, to 2 million pesos, and was pressing federal officials to match those allotments. He later proposed an 8-million-peso cooperative fund for the following year, even as the Secretariat of Finance tried to reduce contributions to Nuevo León's road program by 500,000 pesos annually.[39]

This ambition to further expand the road-building budget created tensions with SCOP over joint funding that boiled over into a

separate dispute with SEOCC over wages. During federal-state negotiations over the proposed budget, the union complained that the
JLCNL was often late when making payroll. In March 1942 agency
supervisors disciplined *peones* who had begun selling their time
cards to pawn shops for cash. The SEOCC came to their defense,
arguing that delays in processing submitted time cards had forced
some workers to wait more than a week to be paid. Although JLC
managers agreed not to fire the individuals involved, they nevertheless refused to permit workers to pawn their time cards, citing the
potential for abuse by third-party lenders.[40]

This issue over wage payments indicated a series of larger potential problems between federal power and Nuevo León's government.
Although both sides ostensibly agreed on the necessity and importance of road building, institutional inertia, wartime austerity, and
the potential for personal acrimony affected how the bureaucracy
operated. Ultimately SCOP and the Finance Ministry were successful in their attempts to restrict Salinas's hoped-for budget increases.
By the fall of 1942 the governor had abandoned the larger plan. In
its place he wrote SCOP again, making special line-item requests in
the next federal expenditure for two motor bridges that remained
unfunded. As the war consumed resources, SCOP tightened control
over federal allocations to JLC budgets.[41]

Veracruz

Public officials took major steps to reform the contentious statewide road-building bureaucracy through further decentralization of
some major activities. In January 1940 they released guidelines that
addressed all work on motor highways, empowering new regional
committees to collect tax revenues and locally direct construction
projects. The structure of these new groups resembled the format
SCOP used for the Juntas Locales de Caminos. An executive board
included representatives for the governor, local chambers of commerce and industry, labor unions, and regional transportation companies, while the state director general for roads—at the time Miguel
Cataño Morlet—appointed an engineer as technical chief. State guidelines required committees to begin collecting local taxes within sev-

enty days of formation, depositing the funds into an account with an area bank.[42]

This reorganization of the state road-building bureaucracy was a capitulation to the persistent administrative problems at the JCCV, which had slowed roadwork and forced bailouts in the 1930s. Veracruz officials decided to give local road committees greater control over construction of the Orizaba-Córdoba and Coatzacoalcos-Minatitlán highways. Under this policy each organization took charge of an individual road-building project, handling day-to-day operations in conjunction with the state director general of roads. The JCCV remained a logistical partner, supplying field crews and equipment to build the routes that these regional boards approved. The arrangement, shifting aspects of road-building management out of the JCCV, appeared successful. Between 1940 and 1943, although the JCCV cycled through eleven different directors—some of whom served only a handful of weeks—Veracruz added 350 kilometers to the state highway network and increased spending for new roads to 2 million pesos annually.[43]

Alongside the greater power of regional committees over motor highways, the state government also enhanced DCOP's authority. In 1940 Luis G. Rendón joined the agency as its new head of the road-building section, having previously gained regional experience as a technical manager at the Jalapa Railroad and Power Company in the 1920s and early 1930s. He served at DCOP until 1943, after which he took over the JCCV for two months before going to work in the private sector as a technical consultant.[44]

During his tenure as DCOP's road-building chief, Rendón centralized decision making for aid requests under the auspices of his office and standardized the application process for communities to obtain materials and equipment for local repairs. For example, in January 1942 Donato Miranda, municipal president of Chicontepec, a county of twelve thousand inhabitants in northwestern Veracruz, wrote an urgent letter to the state government requesting shovels, handcarts, pickaxes, hammers, and supplies to repair the local road. Eight days later Rendón dispatched a succinct reply, informing him that the government could not act until the *municipio* formed separate road and civic improvement boards, composed of "honorable" local residents, to file these requests. Miranda organized the required committees, reapplying in March; this time, with all documentation in order, Rendón dispatched the tools and supplies.[45]

Rendón also ruled on financial disputes that communities lodged against the state and sent inspectors to evaluate outstanding cases. In the town of Jesús Carranza, citizens demanded that the 1-peso duty collected on local beer sales should go to local road repairs and other civic improvements. Community leaders said the monies were instead going to support construction on a road more than 130 kilometers away, which had little benefit to their area. Rendón denied the request and reaffirmed the policy that the road commission in Coatzacoalcos controlled the dispersal of funds in that region. In response the town turned to Victoriano Andrade, a deputy from Veracruz in the federal Congress with a reputation for defending indigenous rights. After some negotiations, Andrade finally reached a compromise with DCOP, which kept the 1-peso tax in place but provided a monthly stipend of 500 pesos from the governor to cover local road repairs in Jesús Carranza.[46]

Under Rendón DCOP established new satellite offices staffed by junior and midcareer engineers who conducted field surveys, inspected locally cleared paths in advance of road construction, and investigated allegations of corruption. In the town of Xico, about an hour south of Xalapa, Guillermo Samaniego, a technical inspector for the agency, uncovered allegations of abuse of power and fraud committed by the local road board president. He also learned that angry farmers had formed an unofficial pro-road committee to raise local funds for a much-needed *camino vecinal*. Samaniego sent a report about these matters to Rendón, who soon revoked the original road committee's credentials and extended DCOP's official recognition to the group run by the farmers.[47]

In December 1944 Adolfo Ruíz Cortines became governor of Veracruz and presided over another transition within the organizational structure of the state road-building bureaucracy. DCOP continued in its capacity as the primary agency involved with evaluating and coordinating local requests for financial aid, materials, and equipment. Possibly as a demotion in its mission, however, state and federal

officials left the JCCV solely in charge of conserving and repairing all national highways in Veracruz. This shift appeared to bring
some stability to the agency's management structure by providing
the organization with a clearer assignment that focused its efforts.
Eusebio Rendón Jarillo, who was appointed JCCV chief in May 1943,
remained in that position for three years, easily one of the longest
terms in the organization's history up to that time.[48]

Local Politics and Environmental Impact

During the early 1940s road-building agencies in both states played
an increasingly active role in regional and local mediation over spatial and environmental disputes. As construction crews enlarged the
highway network and extended new roads to a greater segment of the
population, these efforts also heightened social tensions as affected
individuals and groups responded to the impact of road building and
motor travel. In Córdoba, Veracruz, the pro-highway committee wrote
SCOP about the growing problem of roadside advertising, emphasizing their fear that increased use of billboards detracted from the
region's natural landscape and hurt the burgeoning tourism trade.
The agency made no promises to respond to their request beyond
a simple reminder to the committee to review federal guidelines
on the use of commercial marquees to determine whether owners
were in compliance.[49]

While it is unclear whether Córdoba's residents pressed the issue
further, evidence from Nuevo León indicates that advertisers' over-
use of public space was a palpable concern for local residents and
state officials. JLCNL chief Dominguez ordered companies to remove
advertising erected too close to highways due to the danger posed to
drivers' safety and threatened that the state would do so itself after
the deadline elapsed. This assertive state regulatory agenda extended
to other forms of roadside construction that affected public uses of
space and the natural environment. Although representatives of the
road-building bureaucracy and local communities sought to resolve
outstanding problems, they often clashed over the long-term impact of
construction projects that affected existing land-use arrangements.[50]

In Nuevo León access to water emerged as a serious point of

contention between road engineers and residents who lived along highways. For example, on the route from Monterrey to Saltillo, local citizens dug ditches to carry water to their farms; over time these irrigation networks expanded and came closer to the highway's shoulder. When the JLCNL decided to widen the existing lanes, it claimed that an improved road took priority as a matter of public safety. Domínguez ordered the affected communities to redirect the irrigation works, adding that road crews would fill in any channels if residents failed to comply by a certain date.[51]

Other incidents showed how individuals could successfully extract compensation from an unwilling state agency. Ranchers and farmers whose property was adversely affected by motor traffic petitioned state authorities for redress. In July 1942 Eliseo Sánchez complained to Nuevo León officials that because his land had been divided after a new road was built, his cattle could no longer safely reach their water source. When a motorist struck one of the cows, he renewed his plea for something to be done. The JLCNL finally agreed to pay for and install a pipe to transport water beneath the road, bringing it closer to the herd. In Apodaca, 22 kilometers northwest of Monterrey, José Treviño Sandoval wrote to the state governor, reminding him of a lapsed promise that the JLCNL had made to build irrigation on his property to divert water away from the highway, after a motorist had struck one of his cattle. Salinas's administrative chief, Armando Arteaga y Santoyo, ordered Domínguez to fulfill the pledge.[52]

When road crews damaged local property or a resident accused the government of illegally acquiring land without paying for it, political confrontations threatened to delay roadwork and could force financial settlements. In the summer of 1941, when the JLCNL appropriated a portion of Ernestina Díaz de Leal's land for the highway to Reynosa, she filed a lawsuit against the agency, arguing that it had done so without her approval. While she lost her case against the JLCNL due to a procedural technicality the government exploited, she nevertheless forced a work stoppage on the highway that lasted weeks until a judge ruled on the matter.[53]

Other people who lacked Díaz's means to wage a court battle, however, had fewer options to block work or respond to abuses of power. Petra and Maria Rodríguez were among residents who lost their modest homes when road crews leveled the properties to open space for the highway that connected their town, Villa de China, to the state capital. "We're old women who have no men to help us," they pleaded to the road commission, adding hopefully, "Due to our advanced age and lack of support, we have no doubt you will know how to help us." Municipal officials often declined to reimburse losses for fear of setting a costly precedent. Likewise, in Cadereyta, 40 kilometers east of Monterrey, local authorities wrote to the governor advising him to reject a resident's request for an indemnity. The mayor argued that any financial compensation for the construction of the highway would generate additional claims and strain the city's limited finances.[54]

Landowners in Veracruz became some of the most active critics of state road-building projects. Rodolfo Zamora, a coffee and orange grower from Coatepec, told Governor Cerdán that DCOP's road crews had ruined his crops when they converted a portion of his land into a makeshift quarry to supply paving materials for roadwork. Zamora had his own workers block completed portions of the road until the state promised to reimburse him for the losses incurred. A wealthy landowner in Xico, Dario Soto Peredo, complained to DCOP about a proposed 7-kilometer road extension. The route would have cut through a ranch he owned, which he claimed placed his entire investment in jeopardy. In opposing the project, Soto's lawyers wrote that it would be too dangerous to drivers since other local ranchers had not adequately fenced their properties to prevent cattle from crossing onto the road.[55]

In other instances rural communities complained about poor maintenance of existing roads, which contrasted sharply with the promises by national and state leaders to allocate generous sums for upkeep. In 1942 the route that connected towns in the *municipio* of Doctor Arroyo, Nuevo León, to the Pan-American Highway had deteriorated significantly due to lack of state maintenance. Mountainous terrain and heavy seasonal rains further aggravated the problem and interfered with farmers' ability to transport produce. Public motor travel had become so difficult residents feared a reversal in economic growth and a return to regional isolation. A local civic improvement committee implored the government to repair the motorway, arguing it could mean the difference between "life and death to the area's communities."[56]

Public officials and local communities also clashed over excessive speeding and reckless driving on highways and *caminos vecinales*. State newspapers reported in grisly detail on auto collisions, often involving pedestrian deaths due to driver distraction or intoxication, which stirred significant public safety concerns. By the mid-1940s authorities across Mexico had raised fines against drunk driving and other offenses. They also reengineered roads, widening lanes to reduce the risk of collision from oncoming traffic.[57]

Regional law enforcement, however, remained lax as states favored long-distance motor travel to facilitate commerce and tourism. The results of this policy could be tragic. In one case a bus traveling to Reynosa from Monterrey struck and killed a six-year-old girl walking along the highway that ran through her town. Residents of Paso de la Loma discovered that the driver lacked a proper license and had little experience operating large passenger vehicles. In the subsequent public outcry to the JLCNL, the dead girl's family and neighbors reminded the agency that this accident was not the first they had suffered from state neglect and blamed the problem on reckless driving by motorists passing through the region.[58]

These types of problems eventually pushed state authorities to act. By the mid-1940s federal and state authorities implemented more regulations on road usage, privileging motorists' right-of-way and ordering farmers to keep cattle off highway shoulders. Whereas government ordinances and the courts had protected open access for motor travel, policing efforts restricted other kinds of activities. Citing safety concerns, public officials reiterated ideas about roads as places meant primarily for motor travel.[59]

The Conservative Turn Continues

As World War II came to an end, President Ávila Camacho was reticent to break with business leaders over worker demands for better pay and living standards. For years, while the global conflict carried on, his government had appealed to national unity and the war effort as reasons wages needed to remain low and socioeconomic reforms had to be delayed.[60] After 1945, under pressure from labor groups, national officials ordered the Secretariat of Labor to conduct formal reviews of contentious disputes between workers and management when they erupted. These actions may simply have paid lip service to popular demands for redress of the labor market. Authorities appeared to stall for time, allowing the business community to retaliate against strikers they later labeled communist sympathizers. For example, in April 1946 drivers of several bus and truck companies in Monterrey went on strike demanding higher wages and called on the government to nationalize the firms involved. By August the workers had gained widespread regional support among numerous industrial, film, and agricultural unions whose members sent dozens of letters and telegrams to President Ávila Camacho, urging him to intervene on their behalf.[61]

To further raise awareness some of the drivers went on hunger strike with the plan to travel to Mexico City to press their case. This "hunger caravan," as it became known, reached the capital by the end of the month after passing through numerous towns and cities. In San Luis Potosí the local branch of the CTM wrote of the great "show of sacrifice" that the strikers made against the companies in Nuevo León.[62]

When the workers arrived in Mexico City, their opponents fought back. Martín Ruíz, a leader of the Alianza de Camioneros de México, wrote the president, calling the strike unjustified as it paralyzed transportation in Monterrey. He also claimed that communists had infiltrated the movement: "It is indisputable that these strikes are a typical case of communist penetration and predominance, since the tactics of struggle put into action reveal them to be so; besides, in all of these occasions where an agreement is about to be reached . . . communist representatives have destroyed the possibilities [for a resolution]."[63] Lawyers for the transportation companies also obtained a writ of *amparo* against the drivers from a district judge in Monterrey. Ruíz asked that Ávila Camacho respect the ruling, which called the strike illegal. The unions denounced the judge's decision, writing to the president well into September that the case should go before the Supreme Court.[64]

Under pressure to act Ávila Camacho authorized the Secretariat of Labor to review the case. This decision did little to placate the strikers, who carried their campaign into October. By then the CTM's national executive committee had offered to negotiate a resolution whereby the two sides would hear out the grievances. They proposed the creation of a new drivers' union in Monterrey, which would hire the strikers and allow them to run specific routes in the city and have access to a line of credit provided by the federal government.[65]

In response the leaders of eight transportation companies from Monterrey sent their own proposal to Ávila Camacho. They asked him, first, to speak to Governor de la Garza about blocking an anti-business transportation bill that had passed the state legislature; second, to show support for an increase in urban bus fares; and third, to block the formation of a new driver-controlled local bus cooperative. At the end of October Ávila Camacho's personal secretary, Carlos Guzmán, wrote de la Garza, including a copy of the proposal the companies had sent to the president, but not one of the CTM plan. Guzmán told him the governor that the president wanted him to consider the issues raised by the owners and resolve the matter.[66]

This tactic appears typical of Ávila Camacho's dealings with business leaders and workers. On the one hand, he tried to placate the strikers and labor supporters with calls for formal reviews by the government. On the other hand, he acted more directly in addressing company grievances through backchannels with regional power brokers. This approach allowed the president to appear to mediate while carefully reinforcing the owners' bargaining position. Ávila Camacho did nothing to move against the writ of *amparo*, which the companies had won. Moreover no clear results appear to have come from the formal review that the Secretariat of Labor conducted. Instead the business leaders may have found the president much more responsive in dealing with their concerns through his personal secretary.

For transport workers, Ávila Camacho's mediation over this dispute, and related ones, meant a kind of slow defeat. He gave the appearance of receptivity toward the unions but did little to actually change the status quo that favored company interests. Ultimately this episode embodied the larger trend in the 1940s in favor of capitalist, pro-market policies that gave the benefit of the doubt to businesses and investors while attempting to neutralize labor activism. It was a tactic that the president's successor, Miguel Alemán, would continue.[67]

In September 1946, in his last official state of the union address, Ávila Camacho equated the "progress of a nation" with the "cumulative effect of its public works" and cited road building as a key metric. In the preceding twelve months his administration had overseen more than 14,000 kilometers of road-related projects nationwide and granted almost five thousand new heavy-vehicle permits to private companies in the transportation industry. Federal and state officials, aided by foreign investment, had cooperated in an ongoing program of land surveys, path clearing, paving, and road maintenance. Ávila Camacho saw this work as a means to transform society and improve material conditions for the nation's citizens. His speeches about road building tended toward the dramatic, grasping for the populist support his predecessor had ably commanded: "If schools can liberate us from ignorance, then highways . . . can help to liberate us from misery."[68]

The president's lofty rhetoric about roads, however, did not adequately reflect the reality of carrying out these programs locally. SCOP and the national government took a heavy hand against states that pushed to ambitiously increase spending. Abuses of power occurred as unexplained delays in wage payments forced engineers and managers to cope with angry workers and suspend construction activities. Citizens also complained about abandoned roads, lack of maintenance, illegal land seizures, and damaged property. Wartime black markets charged high prices for scarce goods, leading to public clashes over access to resources. Likewise transportation workers suffered low wages and a lack of labor rights.

Although the outbreak of World War II had improved U.S.-Mexican bilateral relations and buttressed Ávila Camacho's politics of national unity, underlying problems of corruption and inefficiency continued. The postwar era brought large sums of money to Mexico, which enhanced road-building efforts but did not profoundly reverse the political trends of the early 1940s. The following chapter examines how a new president, Miguel Alemán, allocated more spending and built more roads than any of his predecessors but also tolerated more cronyism and backroom deals. He continued Ávila Camacho's close ties with the United States, establishing bilateral developmental programs that invested extensively in new highways against the backdrop of an encroaching cold war with the Soviet Union. Arguably these 1940s policies helped set the foundation for the modern North American economic relationship that had coalesced fully by century's end.

5

"Those Who Do Not Look Forward Are Left Behind"

Alemanismo's Road to Prosperity

A t five o'clock in the afternoon on February 15, 1949, a Catholic priest, Carlos Álvarez, led a procession through Monterrey's new R. E. Olds (REO) Motor Car Company assembly plant, blessing the 10,000-square-meter facility capable of producing 150 motor buses monthly. Joining him in this benediction were powerful and influential members of the local elite alongside U.S. visitors that included John Clarke, president of REO Motors, and Edwin Jones, chief executive of Wells Fargo Bank. President Miguel Alemán and Nuevo León's governor Arturo B. de la Garza sent personal representatives to deliver speeches and inaugurate the factory, while América Domínguez de Garza, widow of Arturo Garza—who had originally brought REO Motors to Monterrey—broke a champagne bottle over the first truck to come off the production line. XEMR, a local radio station, broadcast the entire ceremony live, and *El Porvenir* dedicated a full page of its society section to the event.[1]

During a speech at the gathering Manuel Suárez Mier, general manager of REO's Mexican subsidiary, placed the new facility within the context of national postwar road-building policy: "In the coming years, this country will open more than 100,000 kilometers of new *caminos vecinales*, which will undeniably require efficient and comprehensive transport vehicles." To achieve this end he suggested a private initiative to invest 20 million pesos into regional road-building efforts. He framed this program as part of a broader drive to modernize the nation: "These routes and motor vehicles will carry Progress to all parts of the Republic like blood circulating vigorously through the arteries of a young nation, providing inexhaustible wealth."[2]

Suárez's poetic words captured the zeitgeist of 1940s Mexico as the nation looked forward to enjoying the fruits of postwar industrialization. As early as 1942 people had already begun envisioning the long-term opportunities for economic development that came with joining the United States in World War II. In May *El Norte* had argued that Latin America was on the cusp of a major realignment in trade as Washington looked to the Western Hemisphere as a source for raw materials. In an editorial titled "Después de la guerra: quien no mira adelante, atrás se queda" (After the war: Those who do not look forward are left behind), it urged Mexico and the rest of the region to grasp the shift under way: "Not long ago, the United States imported its wool from Australia, but now it will come from Uruguay, Chile and Argentina; zinc, copper and lead will come from Mexico, Chile and Peru; tin from Bolivia; magnesium—essential for steel production—from Cuba and Brazil. The United States, in turn, will invest credit and provide manufactured goods that Latin America still does not produce. At the same time, this powerful Anglo-Saxon nation will help the rest of the New World in building highways and fomenting its industry."[3] The newspaper described the need to rebuild war-torn Europe as the impetus for greater U.S. reliance on its hemispheric partners. "This will offer significant opportunity, on a grand scale, for new markets that . . . *both Americas* could collaborate in to great advantage."[4]

El Norte saw U.S. banks playing an integral role in driving early postwar economic and infrastructural development. It cited Siegfried Stern, vice president of Chase National Bank, a leading voice in favor of a vision for postwar Pan-Americanism. He highlighted "sustained internal political stability" and "fiscal policies perfectly in accord with the protection of capital, allowing for new U.S. investment and credit in equal cooperation with national partners [in Latin America]."[5]

In 1943 the weekly magazine *El Tiempo* predicted rising postwar living standards and reduced income inequality as rapid U.S.-

style economic modernization transformed the countryside.[6] At a bilateral summit that year in Monterrey, Presidents Manuel Ávila Camacho and Franklin Delano Roosevelt framed wartime bilateral relations as one part of a much larger socioeconomic whole that embraced Pan-Americanism as a basis for postwar development. Speaking of the United States and Mexico as equal partners, Ávila Camacho declared, "Geography has made us a natural bridge for reconciliation between the Latin and Saxon cultures of the continent. If there is a place where the thesis of the Good Neighbor policy can be tested, it is right here in the juxtaposition of these two lands." In his own remarks Roosevelt touched upon themes of modernization that echoed *El Tiempo*'s view: "The grand Mexican family is on a path to greater progress, which will allow all of its members to enjoy security and opportunity. The United States government and my fellow citizens are ready to contribute."[7]

In the summer of 1946, as a presidential candidate for the Partido Revolucionario Institucional (Institutional Revolutionary Party, PRI), Miguel Alemán campaigned on a domestic platform that emphasized road building as a means to spur economic growth in rural communities. This policy represented a return to prewar developmental messages that emphasized market access and motor tourism as key economic drivers. His rhetoric was strongly influenced by the memory of years of economic austerity that had confronted Mexico after nationalization of the oil industry and during the Second World War.[8]

In public speeches to voters he acknowledged that industrial and infrastructural modernization was lacking in rural areas and promised to address this challenge during his term. For Alemán, the question was not whether Mexico would modernize. The contours of that debate had largely been settled after more than two decades of infrastructure development, which occurred alongside new programs that brought in significant private and foreign financial investment. Instead the subject at hand for the PRI candidate was how his government would distribute the gains of economic and industrial growth now that Mexico had emerged from seven years of national austerity.[9]

When Alemán took office in December 1946, SCOP estimated

that only one-fifth of the nation's landmass was connected to a major motorway like the Pan-American and Inter-Oceanic highways. Critics argued that this figure represented too little progress given the scope of the decades-long road construction program. They noted that geographic isolation remained a palpable, everyday challenge for many Mexicans. In response Alemán unveiled a new infrastructure-planning agenda that focused on building asphalt-concrete roads to local communities. Between 1946 and 1952 the rate of growth for the national highway network rose 80 percent to an average of 2,250 kilometers of new roads annually. In 1948, with the creation of the Comisión Nacional de Caminos Vecinales (National Commission for Local Roads, CNCV), which included influential business leaders like Rómulo O'Farrill on its board of directors, Alemán's national policies prioritized construction of regional motorways to boost rural access to the federal highway system.[10]

The president and his associates also expanded the political, economic, and legal framework for road building that Ávila Camacho and earlier presidents had established. Alemán's team employed populist language to frame their case for new roads, citing regional demand for transportation infrastructure as a major factor in planning efforts. The architects of Alemanismo appropriated themes of rural improvement to advance their political and commercial priorities. Maria Antonia Martínez asserts that the president promoted "economic development based on the preeminence of the State as an economic agent. . . . In political terms, the model was rooted in the use of democratic rhetoric in concert with the prevailing international discourses and a practical tendency that favored consolidating greater control over distinct actors."[11]

The national government acknowledged the claims of local communities for greater access to motor routes but asserted that expanding the economy must take precedence. Federal officials threaded together policies that touted commercial developmental needs, while arguing that a majority of citizens would benefit from this strategy. National and state governments looked to add amenable rural and labor groups as subordinate partners in postwar road construction programs.[12]

During the late 1940s and early 1950s Alemán appointed judges who upheld the jurisprudence established under his predecessor. The SCJN remained an ardent protector of the public's right to freely use federal roads. It also maintained the legal distinction that limited state government power while reaffirming the Federal Road Police's jurisdiction to regulate motor traffic. For example, in August 1950 the SCJN ruled that state officials could not restrict bus service on the national highways to Monterrey. The decision, however, proved to be a mixed victory for the plaintiff, Lorenzo Garza, who operated a transportation company that wanted to carry goods as well as people on buses to the state capital. The court agreed that the Federal Road Police had jurisdiction over the highway and thus could regulate the company's activities in the name of public safety.[13]

No major cases that came before the SCJN changed the structure of how Mexico governed land expropriation for road building since the early 1940s. Instead the Supreme Court largely continued to protect regional private interests against state-level political meddling. It also allowed bus companies to choose their own routes without interference from state government regulators. Alemán had ensured the court's ideological transition from a firebrand body that eagerly challenged private power in the 1930s to one that guarded commercial interests in the 1950s.[14]

During this period the national government set in motion road-building policies that would connect nearly 30 percent of the country to the highway system by the end of the next decade. In total, under SCOP and the CNCV this effort included a dense network of regional motorways and nearly two dozen highways that connected the country to its seaports and national borders. As they built new roads, federal and state officials pointed to this expansion as a key metric to advertise Alemán's Mexico as a modernizing nation to domestic and foreign audiences alike.[15]

Important sources of funding for the country's developmental strategy came from postwar public and private investment from the United States. In February 1947, following a bilateral summit in Mexico City, President Harry S. Truman promised to extend to Mexico additional loans via the Export-Import Bank to help build roads.[16] Supplementing these investments, the Bank of America of California gave Alemán's government a 9-million-dollar line of credit to build new motor highways in northern Mexico.[17] *El Porvenir* quoted U.S. Embassy officials championing these efforts as part of a larger, geopolitical approach against the Soviet Union: "A prosperous neighbor is better than a poor one . . . and won't be so easily smitten with Communism."[18]

The national government used U.S. investment and a domestic bond issue worth 100 million pesos, to boost spending on roads well above wartime levels. In 1947 SCOP allocated 150 million pesos for road construction, a 50 percent increase over the 1945 budget. A year later federal authorities doubled road funding to 300 million pesos. SCOP and state agencies oversaw advances in the construction of the Matamoros-Monterrey-Mazatlán Highway, which connected the Gulf coast to the Pacific Ocean, as well as completion of the Pan-American Highway to Guatemala and new northern border highways from Ciudad Juárez, Piedras Negras, and Nogales to Guadalajara and Mexico City. The federal government also funded smaller projects, often through the CNCV, including the 50-kilometer Tampico-Tuxpan Highway and a 64-kilometer route between the ports of Veracruz and Alvarado, which facilitated regional trade and tourism.[19]

Even as the national road system grew dramatically under Alemán, accusations of cronyism, fraud, and abuse of power marred construction efforts. Rural citizens who had been promised new roads later complained of their communities being bypassed in engineering blueprints without any clear explanation by authorities. Following storms that damaged transport infrastructure, others alleged that the government showed favoritism in the decision-making process to allocate funds for road repairs. Although millions of pesos of new investment went into roadwork in the late 1940s and early 1950s, officials wrote about limited resources in response to pleas from communities for help. The best aid that some local road-building boards could hope for from state agencies was secondhand equipment, including old shovels and pickaxes. In other instances community leaders fought one another over control of road-building funds, denouncing crooked deals made by their rivals and appealing to outside authorities to adjudicate matters.[20]

In the face of rumors and allegations, Alemán's political surrogates in the federal Chamber of Deputies defended their patron's road-building plans. After the 1949 presidential address, Armando del Castillo Franco, leader of the Chamber, declared, "In the ongoing work of building highways [and] roads . . . you are achieving, Mr. President, the basic premise of the Mexican Revolution: to unite, not divide, Mexicans in matters spiritual and moral, as well as material." Del Castillo omitted any sense of the messy power politics that actually marked road building and instead characterized it in almost romantic, heroic terms.[21]

Similarly U.S. travelogues about Mexico in the late 1940s and early 1950s conflated the nation's progress with road building. Reporting from Mexico City in 1949 for the *New York Times*, Roland Goodman painted a vibrant portrait of "the roads newly ready for the everyday motorist," drawing on a recognizable narrative of modernity and technological progress to emphasize the social impact of recent highway construction. Whereas old routes had been traversable only on horseback, he wrote that the Pan-American Highway connected "ancient" mountainous settlements more easily "to the outside world."[22]

In the fall of 1952 the *Times* published another of Goodman's travel reports to commemorate the end of Alemán's tenure as president. The story detailed improvements Mexico had made in recent years, building new roads that opened large swathes of territory to U.S. tourists. Alongside mention of the Pan-American Highway, Goodman listed a series of other accomplishments, such as a new airport and a "gigantic" 105,000-seat stadium in the federal capital, which he speculated could signal Olympic aspirations. He also hailed air-conditioned bus service from the border as a welcome improvement.[23]

Alemán not only aggressively sought out foreign investment to drive economic growth; he also looked to road building as a tool to consolidate political power. Construction efforts spread state largesse, bought local allies, and handed out rewards to friends. The tens of millions of pesos that Alemán poured into programs for highways and *caminos vecinales* benefited the ambitions of businessmen linked to the automobile industry. For example, Rómulo O'Farrill, a close friend of Alemán and *prestanombre* (nominal holder) for some of his business interests, owned a Packard car dealership on Avenida Bucareli in Mexico City and a vehicle assembly plant in the state of Puebla.[24] O'Farrill also published the newspaper *Novedades*, which he used to champion motor travel and road construction, touting the government's accomplishments in numerous articles.[25]

The CNCV worked with construction companies linked to O'Farrill and other business elites to create public-private ventures for new roads. These firms relied on the CNCV to obtain federal credit on favorable terms, buy equipment and materials, and begin building regional motorways.[26] The government's enthusiasm for deal making on *caminos vecinales* reflected a larger problem of entrenched cronyism.

Although Alemán's rhetoric held up rural Mexicans as the primary beneficiaries of postwar economic development, in practice new highways and local roads had a much more ambivalent impact on their lives. If farmers happened to profit thanks to a given construction project, the president, and his elite backers in the CNCV, gladly took credit. Nevertheless the federal government's aim remained focused on large-scale commercial development to reshape the national economy. Problems of uneven funding and regional inequality plagued road building under *Alemanismo* as communities in Nuevo León and Veracruz increasingly voiced concern that public officials had failed to fulfill the promises they made for improved transportation infrastructure.

State Road-Building Efforts after 1946

Many private- and public-sector groups in Nuevo León eagerly embraced President Alemán's plan to build roads, developing creative solutions to quickly open new motorways. On 11 December 1946 the new president received a proposal from the Pro-Highway Committee of Monterrey-García, a group made up of leading state politicians, including retired governor Bonifacio Salinas Leal, and members of prominent Monterrey families like the Garzas, Treviños, and Lozanos. The committee's president, Jenaro Garza Sepúlveda, discussed the construction of a new road between Monterrey and Paredón, Coahuila, which would run through Villa de Garcia using a path already cleared and graded by the Ferrocarril Central Mexicano for a rail line it never completed. Garza called the project "easily executable and cost-effective," noting that the president needed only to transfer control of the land from the federal Department for Railroads to scop.[27]

The state government had already completed feasibility studies and the plan enjoyed the support of local landowners. Garza added, "We are sure that this [road] will help drive regional agriculture and ranching, allowing for no less than thirty thousand hectares of land to be cultivated. It will also attract tourism, since there are many picturesque sites along the short route from Monterrey to Villa de Garcia."[28] Almost two weeks later the committee received the president's decision: Alemán agreed to the project, transferring it to scop for execution after December 27.[29]

In the same week that Jenaro Garza made his case for the Monterrey-Garcia-Paredón road, Governor Arturo B. de la Garza delivered a separate report to President Alemán in support of a new central highway from the border at Piedras Negras, Coahuila, to Mexico City. The plan had emerged out of a broad working group composed of the state governors of Nuevo León, Coahuila, Tamaulipas, San Luis Potosí, Guanajuato, Jalisco, and Querétaro. They proposed construction of a series of extensions to the existing Pan-American and Inter-Oceanic highways, opening a new route to the United States and also improving motor travel options between the cities of Matamoros, Monterrey, Saltillo, Matehuala (San Luis Potosí), Guadalajara (Jalisco), and León (Guanajuato). In total the plan called for more than 1,000 kilometers of new roads across northern and central Mexico and suggested raising the gasoline tax by 2 cents to pay for the work.[30]

When Alemán authorized this road project soon afterward, it represented the culmination of years of careful planning by public and private supporters. In 1944 the state Chambers of Commerce and Industry from Guanajuato, Coahuila, Nuevo León, Tamaulipas, and San Luis Potosí, in conjunction with the National Association for the Lions Club, had submitted a similar proposal to President Ávila Camacho. It arrived the same day that Jose Antonio de la Lama y Rojas, an artillery lieutenant and right-wing extremist, made an attempt on the president's life. Amid the chaos generated by a failed assassination attempt in wartime, the request was likely overlooked and languished until the governors' working group revived it after 1945. In a letter dated 17 December 1946, the Club de Leones de San Luis Potosí echoed many of Jenaro Garza's sentiments in favor of the Monterrey-Garcia road when it thanked President Alemán for approving the Piedras Negras–Mexico City Highway plan:

> For more than eight years, this chamber of commerce has repre-
> sented the economic needs of the state, calling for the construction
> of a federal and international highway that unites the cities of Pie-
> dras Negras, Saltillo, San Luis Potosí, San Luis de la Paz, Querétaro,
> and Mexico City. It will provide inestimable commercial and indus-
> trial benefits to the Republic, putting the major consumer and dis-
> tribution centers of the capital in direct and easy contact with all
> of the communities that extend along the border with the United
> States of America. It will provide an outlet for all of the goods pro-
> duced in this part of the country, which had long lacked transport
> options to larger consumer markets.[31]

Once approved, the Piedras Negras–Mexico City Highway required effective regional coordination between state governments over the financing extended for construction. In 1947 and 1948 complications reminiscent of challenges during the war erupted between Nuevo León and Tamaulipas over budget allocations for the highway. At the time the Junta Local de Caminos de Nuevo León reported that it had worked with federal authorities to jointly invest 5.9 million pesos, roughly 65 percent of the regional project's estimated cost. Governor de la Garza complained that Tamaulipas had only con-tributed 1.5 million pesos, leaving a 17 percent deficit in the bud-get. Alemán resolved the issue—per de la Garza's suggestion—by extending a 2-million-peso road bond to Tamaulipas, covering the remainder of the expenses.[32]

In 1949, when Ignacio Morones Prieto succeeded Arturo de la Garza as governor of Nuevo León, the Monterrey-Garcia and Piedras Negras–Mexico City routes became parts of a larger six-year plan for road building in the state. The JLCNL listed nine new routes that drew funds from the federal-state cooperative budget and *camino vecinal* construction programs, representing a combined cost of 43 millions pesos. In addition the agency estimated state conservation expenditures for these routes to equal 2.25 million pesos per annum. In total Morones Prieto planned to build almost 2,000 kilometers of new roads in Nuevo León, drawing on a combination of federal and state bond subsidies, plus tax revenues from petroleum and tourism. Between 1945 and 1948 contributions Nuevo León received via the federal gasoline tax increased roughly 11 percent per year, from 1.8 million to 2.5 million pesos. At this rate of growth state officials could expect contributions to exceed 3 million pesos by 1950 and reach 5.5 million by 1955—more than half of the estimated 9.2 million annual budget for road building.[33]

The JLCNL proceeded with construction efforts, using 735 kilometers of national highways in the state as the foundation to expand Nuevo León's network of *caminos vecinales*. By the early 1950s, in conjunction with the CNCV, it had extended regional access by 367 kilometers, building six new local motorways: Sabinas-Parás–General Treviño, Saltillo-García-Monterrey, Monterrey–San Miguel, Cadereyta-Allende, Galeana-Linares, and Villa China–Terán. This roadwork represented more than just the creation of simple dirt or macadam routes in rural areas; it included the construction of mostly asphalt-concrete roads, which were supported by drainage systems to reduce the chance of flooding, and also comprised the building of a least half a dozen motor bridges. Moreover the agency highlighted three future asphalt routes, which would pass through thirteen southern and western counties in the state.[34]

Local enthusiasm remained strong throughout this period. In 1949 the municipal president of Agualeguas, Eleuterio Salinas, wrote to Governor Morones Prieto about the road to Parás: "Our communities are prepared to help the government carry out this work. Although what we can offer is very small we have organized committees that will allow all of our residents to participate according to their economic situation."[35]

In many cases regional coalitions included chambers of commerce and civic associations. They helped to coordinate area labor and make formal requests to state and federal officials to provide greater financial support to local construction projects. For example, in the *municipio* of Sabinas Hidalgo, which was located at the opposite end of the road that ran through Agualeguas from Parás, Margarito Raymundo Salinas led planning efforts. A local commercial and agricultural agent, he enjoyed financial ties with the Mercantile Bank of Monterrey (today Banorte) and owned properties that contained five thousand head of cattle. Salinas offered all of his "moral, material, and economic" cooperation for the road, which would not only serve as a key *camino vecinal* for the state but also would do much to provide greater market access for his goods and those of the rest of the regional community.[36]

This drive to connect once-isolated populations to the state highway network could also devolve into acrimonious local political disputes. As rival groups sought to assert their own vision for regional road-building plans, there existed the possibility for lawsuits and lengthy court battles to slow construction efforts. The legal disputes over the planning for a road from the Pan-American Highway to Villa de Santiago serves as an illustrative case study of this problem. In the fall of 1948 Melquíades Sanchez, the local municipal president, wrote Governor de la Garza, denouncing an area landowner who had opposed the proposed routed: "Alvino Valdés has been extorting the men involved in this project since the beginning, confusing some farmers about his true intent. He is an enemy of progress to this region."

The *camino vecinal* in question would connect a dozen rural communities to Monterrey, 38 kilometers away, and give residents easier access to the Pan-American Highway. In 1947 Pedro Reyna, local boss of the Grupo de Fruteros Nacionales, had organized a pro-road committee to coordinate local support for the construction project. As the plan developed, Alvino Valdés began a campaign in opposition to it, convincing small-scale cultivators in the isolated and mountainous territories west of Santiago to deny monies for a local tax that supported the construction plan. He alleged that members of the pro-highway committee in the village were corrupt and regularly misappropriated the collected funds for their own personal use.[37] Reyna's road committee subsequently wrote to the governor defending itself, arguing that Valdés had impugned their reputation and requesting that a local judge order the release of withheld tax funds.[38]

In 1949, over eight months after the dispute began, Valdés brought another lawsuit before the court in Cadereyta, forcing a work stoppage. He argued against completion of the route through the mountains to Santiago at Cola de Caballo, a nearby waterfall and potential tourism destination. Instead Valdés advocated for an alternative path, with its terminus at Cañón de San Francisco, farther north. A local contractor, Diego Saldívar, complicated matters when he offered to build this other road for less than the already agreed upon budget for the Cola de Caballo project awarded to the construction firm Caminos Desmontes, S.A. Despite a survey by the JLCNL attesting to the technical impracticality of building through Cañón de San Francisco, state officials in charge of overseeing the funds canceled the more expensive deal and gave Valdés and Saldívar their backing.[39]

This change of course created new delays as the entire construction schedule stretched deeper into 1949. First, Saldívar repeatedly missed deadlines to present a working plan for his project to the JLCNL. Then Reyna's pro-highway committee in Santiago discovered that the terminus for this alternate road ran alongside commercial property Valdés owned. They accused him of trying to benefit financially at the expense of providing road access to a greater part of the region's population. As a result the Grupo de Fruteros Nacionales brought a countersuit against Valdés in Cadereyta, ordering him to stop opposition to the road tax meant to fund the project and to personally reimburse the amount that had been withheld from revenue collectors. Reyna further implored the governor to dismiss Saldívar's contract as an "imaginary project" and restore the agreement with the original company to finish the road at Cola de Caballo.[40]

In December 1949, two weeks after Reyna delivered his second complaint, the new governor, Ignacio Morones Prieto, finally backed the original road plan. Following more than a year of personal acrimony and public debate, advocates for the first project linking rural western communities to Santiago gained the needed political endorsements to see it to completion. Construction crews broke ground in the spring of 1950; in May of that year *El Norte* reported that the state government had invested 150,000 pesos in the project and that the Grupo de Fruteros Nacionales contributed an additional 80,000 pesos for work expected to take the rest of the summer to complete. Once finished, the newspaper predicted, "Villa de Santiago will become an important tourist destination. Hidden within this *municipio* is great natural beauty, once almost inaccessible and unknown to the public. Thanks to new roads, these places . . . will be in reach for visitors to experience."[41]

Although the governor had played a key role in deciding the fate of the route, local actors clearly retained considerable agency in developing and contesting construction plans. It is difficult to ascertain which group employed more dirty tricks or insider connections to their advantages, but the Santiago case indicates that much was at stake for local business leaders who hoped to benefit from the national government's program for *caminos vecinales*. They actively engaged government officials as well as the press, insisting on their agendas in public to ensure their voices were heard on matters of regional economic development.

Veracruz

Meanwhile, in Veracruz, institutional dysfunction and challenging environmental conditions continued to plague the road-building bureaucracy. In the spring of 1946, when candidate Alemán had campaigned across the state, he acknowledged the difficulties of regional motor travel due to the poor quality of its many existing *caminos vecinales*. In Tantoyuca, a *municipio* in northern Veracruz, he promised the farmers who had gathered to listen to his speech that he would address local challenges to motor mobility. For years residents had complained that the region's roads were largely impassable for motorbuses during the rainy spring months. Alemán also declared a commitment to finish the Tuxpan-Tampico Highway, which ran through the Huasteca Veracruzana to the state border with Tamaulipas. Pro-road groups saw the project as a key driver for economic development, since it linked to important regional ports and supported a network of roads that added access for interior *municipios* to the state's coastal highway.[42]

Six months after the election it appeared to many local constituents that the project was not proceeding as they had hoped. In a letter dated January 1947, they reminded the new president of his pledge: "This region is the very heart of the Huasteca Veracruzana, the cradle of the Revolution. Sadly, we have learned that the plan for the highway has changed completely. . . . As someone who knows this area and its problems . . . we ask that you intervene on our behalf."[43] The new route bypassed Cerro Azul and six other isolated communities clustered in a densely forested and mountainous part of the Huasteca Baja. Road engineers had made the decision to reroute construction because of the logistical difficulty of building in the area and the potential for delays as a result. Instead of moving forward with the original road through Tantoyuca, SCOP, the CNCV, and state authorities preferred the construction of a series of *caminos vecinales* that linked the county seats of the interior *municipios*, including Tempoal, Panuco, and Chicontepec. This proposal created a junction of motor roads in northern Veracruz between the Tampico-Tuxpan and Perote-Xalapa-Coatepec highways.[44]

A divide existed between political promises and the official policies actually carried out. Federal and state authorities prioritized construction of new roads that enhanced connections to coastal ports and industrial centers above popular concerns for equitable distribution of public spending on transport infrastructure. Even as Alemán touted the work of the CNCV, local leaders in northern Veracruz filed complaints with the state government. They wrote about repeated shortages of materials and equipment that forced communities to share much-needed resources. DCOP officials responded, saying that limited budgets allowed them to do little more besides triage the cases.[45]

These problems in the Huasteca Veracruzana were emblematic

of the conflict between everyday needs and national policy during Alemán's term. On the one hand, much of the political rhetoric promoted state and local needs within the program for *caminos vecinales*. On the other hand, the optimistic language of this period obfuscated the underlying political tensions at the regional level that had marked road-building efforts since the 1920s. Not everyone could benefit equally from the investments being made. Towns with greater political cachet won out over others, despite the president's promises of aid to economically and culturally isolated communities.[46]

From 1944 to 1950 state leadership, first under Governor Adolfo Ruiz Cortines, and later Ángel Carvajal Bernal, enacted three major reforms of the road-building bureaucracy. First, public officials more clearly defined the role of the Junta Central de Caminos de Veracruz, putting it in charge of maintenance and repair of seven branches of the national highway network in the state. The agency also began coordinating more closely with DCOP on construction of new federal highways, providing survey and logistical support for the Tuxpan-Tampico, Conejos-Huatusco, and Córdoba-Veracruz roads, among others. Working through the JCCV, the federal government set up research laboratories to test local soil conditions and construction materials. These facilities experimented with ways to build better roads that could withstand the state's heavy seasonal rains.[47]

Second, the state government appears to have radically consolidated road construction efforts. It reduced the number of agencies involved in the work, while enhancing the authority of DCOP and the JCCV. Official documents from this period make no mention of the Junta Local de Caminos de Orizaba or the independent committees like Coatzacoalcos that had enjoyed greater organizational autonomy in the late 1930s and early 1940s. Instead it appears that federal and state authorities had divided responsibilities between the JCCV and DCOP. For example, both of these agencies worked together on expanding the Córdoba-Orizaba-Veracruz Highway with no reference to the once critical JLC de Orizaba.[48]

Whereas the JCCV remained focused on maintenance and construction of some federal highways, DCOP took charge of all *caminos vecinales*. This decision gave the agency significant influence in how planning and implementation of local roads occurred across the state. In many ways it represented an extension of the duties it carried out in previous years, but now with greater scope thanks to increased federal support for regional motorways. In addition DCOP supervised and enforced state contracts with private companies brought in to supply technical assistance and manual labor. It managed delivery of heavy equipment and motor vehicles to job sites, leasing items like trucks and tractors to private contractors.[49]

Third, the expanded role for private contractors marked one of the biggest changes in Veracruz's road-building bureaucracy. Engineers who had formerly worked for the state government created their own private firms to provide technical support to DCOP. For example, in October 1946 Luis G. Rendón and two other engineers, including his son, Luis Rendón Jr., presented a formal letter of solicitation to agency chief Gustavo Rocha. They proposed carrying out advance surveys of a series of proposed roads in central and southeastern Veracruz that would expand the established Coatzacoalcos-Alvarado Highway built during the war. Rendón and his son estimated a weekly budget of 810 pesos to pay for technical staff and *peones* in four separate survey and construction brigades.[50]

DCOP received competing bids for contract work from private groups from across the country. In July 1946 the state government signed a lucrative deal with Carlos Hernández y Hernández, a Mexico City–based engineer who ran a private contracting firm that built motor roads and bridges. Following CNCV guidelines that emphasized public-private initiatives, Rocha put the company in charge of implementing blueprints for *caminos vecinales* in the state; a federal road bond subsidized construction efforts. In 1948, when Hernández renewed his deal with the state, DCOP officials wrote a nineteen-page contract that dictated specific terms for the relationship, from construction timelines and requirements to setting price levels for labor and materials charges. The agency also required Hernández to make all purchases through official channels in Veracruz, prohibiting him from using outside vendors.[51]

The construction of the Conejos-Huatusco road exemplified how each of these different aspects of the new state road-building bureaucracy came together. Started in 1948, the road ran 95 kilometers from central Veracruz to the Gulf coast, crossing the Perote-Xalapa-Veracruz Highway via the *municipio* of Puente Nacional. Throughout the project engineers faced challenging terrain and weather. The western terminus of the route was in a mountainous region of the state 1,300 meters above sea level with forty-one inches of annual rainfall. It then descended through thick evergreen forests to the coastal *municipio* of Úrsulo Galván, 20 meters above sea level, with a humid tropical climate that received thirty-nine inches of yearly precipitation. Technical crews relied on the federal laboratory in Xalapa to conduct soil analyses, which determined that the land was suitable for construction. In its current condition it could support an initial macadam-gravel path ahead of the application of a more permanent layer of asphalt-concrete.[52]

DCOP assigned Hernández's engineering firm to build the road. It set the budget for the project at 1 million pesos, which funded a construction brigade that employed up to thirty-four laborers, who worked eight- to twelve-hour shifts, typically six days a week, with average weekly payroll expenses of 1,200 pesos. Hernández leased vehicles through the state government to move workers to the job site, paying 6 pesos a day for the equipment. In addition DCOP assigned Paulino Ceballos as a supervising engineer to travel with the contractors and provide regular progress reports to Director Rocha. Work proceeded from the early summer months and into the fall of 1948, with much of the road completed by the end of that year.[53]

One of the principal complaints among local residents had nothing to do with the planned route but rather with the behavior of the road workers. In their off-hours the men visited nearby communities and caused problems at local night spots. In September, Rocha wrote Ceballos ordering him to address this problem: "It has come to my attention that workers on the Conejos-Huatusco highway are frequently causing local disturbances, generating scandal that reflects poorly on this agency. You must ensure they cease their disorderly behavior after hours, reminding them that you are authorized to discipline any personnel involved."[54]

Rocha's letter provides further indications of the notable changes

to the everyday characteristics of the state road-building bureaucracy and its work. DCOP had not only allowed private contractors
to form and direct construction brigades, but this personnel was
being sourced regionally instead of recruited from the communities themselves via the *faena*. This case suggests that eroding communal bonds were to blame as rural towns lost local control over
labor in favor of formal wages and hiring practices supervised by the
state. For this reason road workers may have felt less bound to the
social obligations of the communities where they operated and thus
acted out in areas where they did not suffer punishment from family or friends. Moreover since local leaders wrote DCOP to complain
about this bad behavior, it indicated that typical informal channels
of communication between neighbors to resolve disorderly conduct
were not available options here.

Although state road-building policy at this time favored centralization of its operations, it appears to have had little impact on the
standardization of worker pay. From the mid-1940s onward, daily
wages that the JCCV, DCOP, and private contractors paid to *peones*
differed widely. For example, depending on which part of the state
an agency hired workers, it paid between 1.75 to 4 pesos for the
same kind of work. Other factors were likely at play. Unlike Nuevo
León's state road workers' union, no single union negotiated labor
wages and benefits with state officials in Veracruz. Given the his-
toric rivalry that existed between DCOP and the JCCV, as well as the
presence of numerous private contractors, market demand for labor
may have outstripped its supply.[55]

Amid these institutional changes, Ángel Carvajal Bernal's brief
two-year term as interim governor proved to be one of the most pro-
ductive times for road building in Veracruz. By 1949 the state had
invested 4.5 million in the construction of more than 720 kilometers
of new roads, with well over half of these routes being fully paved
with asphalt-concrete.[56] The following year, when he resigned to
become Alemán's energy secretary, Carvajal Bernal orchestrated an
elaborate farewell tour of the communities that had benefited from
his budgetary largesse. In April, when he arrived in Huatusco, *El Dictamen* described how "campesinos from all of the surrounding com-

munities lined two kilometers of the road into the city to applaud" as the motor caravan passed. Later that day, at a banquet honoring the governor, Dr. Vinicio Méndez gave a toast that highlighted the economic impact of the Huatusco-Conejos Highway: "The work of this highway stands out as essential in uniting Huatusco with all of the principal populations of the state, providing opportunities to the city and the surrounding communities. It brings them closer together with all of Veracruz in a manner both physical and spiritual."[57]

Bad Roads, Broken Promises

As reflected in Dr. Méndez's speech, road building underscored Mexico's developmentalist ambitions after the Revolution. New motor highways and *caminos vecinales* played a key practical and logistical role, providing improved access to regional markets and connecting the nation more closely with foreign economic partners like the United States. They also existed as a kind of public stage to display modernization efforts to national and international audiences. Political and commercial leaders rarely shied from the opportunity to characterize roads as tangible expressions of their vision for building the nation.[58]

Newspapers played an important role as cheerleaders of this work. For example, the editors of *Novedades* made road building and motor tourism fixtures of its daily reporting. In 1948 this coverage increased even more after Rómulo O'Farrill Sr., the owner, also became the general manager after Jorge Pasquel's abrupt departure. Under O'Farrill's leadership the newspaper added a regular section on the progress of regional infrastructure development, including roads. *Novedades* also frequently ran full-page features about tourist destinations across Mexico, with bus and travel advertisements bordering the text.[59]

The positive coverage of motor travel complemented O'Farrill's commercial interests. Not only did he own automobile distributors and an assembly plant; he was also president of the Asociación Mexicana de Caminos (Mexican Road Association, AMC). Founded in 1948, the AMC lobbied heavily on behalf of road construction, working closely with Alemán's government. It joined with other organizations, including the Asociación Mexicana Automovilística (Mexican Automotive Association) and Mexico City's Rotary Club, to promote motor tourism and oppose tax increases on automobile ownership. In turn, *Novedades* transformed the public activities of these groups into front-page stories with fawning media coverage.[60]

Bad roads, motor accidents, and poor maintenance efforts threatened to undermine this narrative of modernity that O'Farrill and other elites preached. Highways and *caminos vecinales* were not simply static monuments to progress; they were dynamic spaces that played an integral role in everyday life. If not properly maintained, they atrophied with usage and weather.

Heavy rains and flooding brought safety concerns to the forefront, igniting popular debate over whether government officials were doing enough to address public welfare. For example, on the morning of 10 October 1950, Hurricane Item made landfall in Veracruz, inflicting significant damage to the highways and *caminos vecinales* that ran between the ports of Los Tuxtlas, Alvarado, and Veracruz, as well as the roads inland to Xalapa and Córdoba. Federal and state authorities responded slowly, and the issue was compounded by what *El Dictamen* described as years of poor maintenance that had left many roads along the coast around San Andrés Tuxtla in poor condition. Enrique Llorente and Juan Cañedo, federal engineers assigned to Xalapa, claimed that media reports had exaggerated the problem. They argued that "at no moment have normal traffic flows been suspended along these routes." Nevertheless local drivers complained of long travel delays due to routes pockmarked with potholes, pools of standing water, and large quantities of mud.[61]

Only after President Alemán announced plans to visit Veracruz in December 1950 did public officials appear to redouble efforts to manage the crisis. In mid-November *El Dictamen* reported that the JCCV had activated more than one hundred service trucks to move materials and workers to key sites along the president's expected travel route from the state capital to the Gulf coast. Local communities criticized this policy, arguing that the state had failed to adequately address the damage wrought by the hurricane. They called the existing conditions around Los Tuxtlas "disgraceful" due to insufficient maintenance policies that had "abandoned roads once completed, leaving them only minimally traversable" when bad weather struck. They also alleged that public authorities were showing favoritism to some communities over others. In the face of this public reproach, road-building officials again defended their work, stating that limited resources forced them to prioritize repair efforts. By early December crews had finished cleanup of the Xalapa-Veracruz and Veracruz-Alvarado–Los Tuxtla highways. Other roads not used during President Alemán's visit were not fully cleared and patched until the end of January 1951.[62]

These problems belied the discourse of progress so popular among national and state leaders. Broken roads, due to bad weather and poor maintenance, left whole areas inaccessible to motor travel. They became tangible examples of a counternarrative about modernity that inverted the symbolism of transportation infrastructure. Media accounts and popular opinion began to ascribe notions of backwardness to motorways in disrepair, reinforcing negative stereotypes of regions that lacked the resources to sustain acceptable and accessible roads. Moreover the decision to triage state responses to problems that emerged due to natural disasters ignited popular resentment with allegations of favoritism and cronyism. It pitted communities against one another as they fought for better access to existing emergency aid for road repairs.

To complicate matters further, these concerns arose at a time of sharp rises in motor travel and foreign tourism in Mexico. In 1946 the federal government logged a record number of motor vehicles in circulation on the nation's roads: 120,906 automobiles, 12,915 motorbuses, and 71,613 cargo trucks. That same year more than 255,000 tourists entered Mexico, most arriving from the U.S.-Mexico border, marking a record high since federal officials began collecting statistical data in 1929. In the early 1950s growth continued apace. Tourism rose roughly 50 percent, with an average 435,000 foreign visitors arriving per year, while the number of motor vehicles in circulation increased to more than 402,000, on average, annually. This raw data translated into swelling rates of motor traffic for the nation's highway network that brought significant popular pressure to make sure routes remained open.[63]

In many respects the problem of road quality reflected some of the political limitations facing the nation. Although road-building campaigns had extended highway access to more areas, this work was far from uniform. Public officials and members of the business community, who saw new roads and motor travel as key drivers of economic growth, had pressed for an "all of the above" strategy. Construction efforts through the CNCV, in coordination with state agencies like the JLCNL and DCOP, prioritized roads that linked economically important regions. Cronyism and incompetence, however, took a toll on these operations. Maintenance budgets were woefully underfunded and failed to cope with the damage wrought by natural disasters. The promises of infrastructure development had crashed into deeper challenges stemming from a lack of political reform. Bad roads that stymied motor travel revealed these underlying problems of inefficiency and mismanagement.

When Alemán started in the presidency, Mexico was at the beginning of a period of great optimism. After years of economic austerity, the late 1940s represented a time where the country hoped to capitalize on the promises of industrialization and economic modernization that federal and state politicians had long made. Building roads fit into this calculus as a policy to facilitate market growth by increasing regional access to motor travel. SCOP and state road-building agencies paved thousands of kilometers of asphalt-coated thoroughfares thanks to a combination of heavy public spending and private investment from domestic and foreign creditors.

Nevertheless the broad promises made on the 1946 campaign trail did not live up to popular expectations. The injection of new investment created more highways and *caminos vecinales* but also divided communities over where these roads should be built. And the problem of corruption was significant. It appears likely that large sums of money were siphoned off through cronyism, graft, and mismanagement during the 1940s and early 1950s. Even as the national government seemed awash in road-building funds, many rural communities were left without adequate resources to repair broken roads.

Alemán, whom Enrique Krauze has called the "businessman president," could not stave off the financial consequences of his decisions favoring wealthy and powerful friends. By 1952, as he prepared to leave office, the country had slumped into an economic crisis. His favorability rating among citizens was so low, due to corruption allegations and the 1948 peso devaluation, that his PRI successor, Adolfo Ruiz Cortines, publicly repudiated much of his political legacy.[64]

Attempting to modernize the nation through road building and the promotion of motor travel could go only so far without sincere political reform. Construction efforts confronted the old problems of corruption and mismanagement, which belied the optimistic pronouncements Alemán and his coterie made about the country's material progress. Despite public outcries, however, this narrative of infrastructure development to promote economic growth ahead of addressing issues of social justice became a familiar one for Mexico as the twentieth century continued.

6

Charting the Contours

State Power in Mexico's Road-Building Efforts

In the 1950s, when drivers planned a shopping trip to Laredo or a weekend getaway to the beach in Veracruz, they may have purchased a highway guide in one of the travel stations that catered to motorists. The *Guía automovilista de carreteras, General Motors' Guide to Mexico's Highways*, and the PEMEX *Guía de carreteras* included maps that depicted the principal highways that crossed the nation with bilingual travel listings for hotels, restaurants, gas stations, and other points of interest.

The guides presented an uncluttered image of Mexico. Bold lines marked out the roads that linked major towns and cities, while drawings of automobiles, tour buses, and airplanes decorated the pages and evoked notions of modernity tied to motor travel. The visual quality of these documents contributed to the dominant narrative that government officials and business people promoted, rendering the political and technical challenges of road building invisible and envisioning highways primarily as conveyors of progress.[1]

Of course roads were not simply physical paths depicted as steadily drawn lines on maps. What the unsuspecting traveler could not have gleaned from these documents were the fights between state officials and property owners over where to build roads, labor tension between engineers and road crews, or the other budgetary and legal squabbles that defined construction efforts. This book has shown that road building was integral to social, political, legal, and economic trends in Mexican society after the Revolution, when people debated the merits and consequences of national modernization.

Between 1917 and 1952 the Mexican government built thousands of kilometers of federal highways and local roads. It channeled millions of pesos of investment into the development of regional transportation infrastructure projects. New roadwork established a foundation for economic growth that connected faraway markets and reduced the cost of travel for average citizens, foreign visitors, and private businesses. It served as an important activity that improved ties with the United States and promoted industrial growth.

The road network that emerged from this work physically underscored notable differences in Nuevo León's and Veracruz's regional economies and populations. It also reflected how these states pursued distinct approaches to the formation of their road-building bureaucracies. In Nuevo León major highways concentrated around Monterrey, like arteries connected to the heart, extending out from the state capital, running northward to the border and southward to Mexico City. From these trunk lines, *caminos vecinales* fanned out to connect the state's much smaller regional cities and rural towns. In contrast, Veracruz's web of highways was much more complex and dense; a federal highway ran between Veracruz and Xalapa en route to Mexico City, but other major thoroughfares connected Tuxpan in the north and the sister cities of Orizaba and Córdoba in the center of the state to Puebla and the national capital. Another important highway ran along the Gulf coast like a spine, connecting all of Veracruz's port cities.

The large amount of money invested invited abuse. Conflicts of interest, insider deals, cronyism, and graft marked road-building efforts. These activities, and other forms of corruption, affected all levels of government and society. From local backroom deals to federal authorities handing out contracts to the companies they owned, individuals and groups looked for advantages to be gained by trying to influence the road-building bureaucracy. In general terms, perhaps three of the national leaders most emblematic of this system of corruption were General Almazán (insider deals), Maximino Ávila Camacho (abuse of power), and President Alemán (cronyism). The first used his public-sector connections to directly benefit his own construction companies; the second eliminated all rivals to his authority in SCOP, replacing them with close allies from Puebla, and also whipped governors into line politically through targeted use of the mechanism to disperse state road-building funds; the third poured millions of pesos into construction deals that benefited the commercial interests of his friends. By evincing a willingness to take advantage of the political and bureaucratic institutions they presided over for personal gain, these national officials set an example that many state and local leaders followed.

Although it is difficult to ascertain how much money corruption diverted from road construction and maintenance, the historical record provides some indication. For example, in the 1940s, as the federal government invested hundreds of millions of pesos into work projects, building new highways and *caminos vecinales*, citizen groups complained about a lack of access to financial resources and equipment. At best, some rural communities could only hope for second-hand tools state agencies finally delivered after repeated requests. Other people alleged favoritism in how the government deployed resources to repair road infrastructure after storms or claimed that promised motor routes had bypassed their town in favor of others.

Control of road-building projects also ignited fierce battles between rival commercial and political groups. In the Amasco case, U.S. investors looking to make inroads through President Ortiz Rubio were outmaneuvered by Almazán's superior political connections. Local fights erupted over who controlled the planning of new motorways and who would benefit economically from construction efforts. In the example from Villa de Santiago, Nuevo León, local businessmen waged a months-long acrimonious dispute over the proposed road to Monterrey, which finally required the courts, the governor, and the state road-building agency to intervene and settle.

The politics (and antics) of the people and organizations that built roads in Mexico underscored the potential for rancor in large modernizing programs. Although the highway system became an important symbol of progress, put on display for national and foreign audiences, beneath the gleaming surface one finds a dynamic, robust, and exceedingly contentious political process. In this context Mexico's history of road building highlights how state formation was as much about managing and coping with the myriad everyday demands, legal concerns, and special interests of the country's citizenry as it was about consolidating national political power after the Revolution.

Remapping National Politics

Beginning in the 1920s the Mexican state used road building to rebalance its political activities with foreign powers and other national, state, and local entities. Plutarco Elías Calles created the National Road Commission as a means to reexert national sovereignty at a time when the United States and Great Britain, through the banks and energy sector, restricted Mexico's economic agency. By fashioning road policies and laws that emphasized the importance of hiring Mexican workers and guaranteed public access to motorways, the Calles government carved out space for the country to exercise political power against foreign influence.

State governments and local committees followed, implementing related strategies. They created organizations that channeled funds into road building and legally challenged entrenched foreign actors, such as the oil companies, over the freedom of movement on privately built motorways. These actions contributed to social movements that imbued road building with a strong sense of nationalism. Federal and state leaders, as well as local groups, cited revolutionary ideals and the need to empower Mexico economically and politically when they discussed the importance of new motorways for the country.

This balancing act took on different forms and gave politicians room to maneuver in terms of rhetoric and policy.[2] For instance, Lázaro Cárdenas pursued pragmatic strategies that built on programs preceding administrations had established to build and finance roads. As Benjamin Fulwider has noted, Cárdenas emphasized the corporatist benefits of road building for *ejidatarios* and other popular constituents, while also pursuing construction of transportation infrastructure that benefited Mexico's commercial sector.[3] The financing of new roadwork, and support for private contractors, allowed President Cárdenas to build alliances with some right-wing politicians and business groups. In 1937 his decision to expand the road bond program initiated by President Abelardo L. Rodríguez generated millions of pesos for highway construction. The groundwork he laid for greater involvement of private investment in road building incidentally served as the framework for future U.S. participation in financing Mexican infrastructure programs in the 1940s and 1950s.

Where Cárdenas left a progressive political mark on road building was in the appointments he made to the Supreme Court. The judges made sweeping decisions against property owners and private companies in cases dealing with land expropriation and road access. The court became a strong protector of state power to acquire land for road building, repeatedly denying plaintiffs' writs of *amparo* to stop construction efforts. Cárdenas's ideological contours were most notable in the SCJN's decisions against the Huasteca Petroleum Company and other powerful foreign-owned oil firms. In 1938, weeks before the president nationalized the industry, the court had denied the appeals these companies filed to restrict public access to the roads they built to facilitate industrial operations.

Mexican presidents played a double game with road-building policy. Where Cárdenas cultivated left-wing and right-wing alliances through the financing and implementation of construction efforts, Manuel Ávila Camacho plotted a similar strategy, but with notably different ends. He appointed Supreme Court justices who took a much more skeptical stance toward state land expropriation, protecting private property rights in road-building cases. With the United States, negotiations to repair diplomatic relations opened the way for millions of dollars in public and private foreign investment in Mexico's highway system. Ávila Camacho used road policy as an integral part of his strategy for national reconciliation. He signaled to Mexican conservatives and foreigners that the federal government had pivoted to the right in political terms, while retaining populist language in speeches that described motorways as "public" goods.

Miguel Alemán went even further than Ávila Camacho in favor of pro-business goals for road building. The "businessman president" pursued economic growth at the expense of workers and farmers but used the idea of road building to articulate a developmentalist message inclusive of organized labor and campesino groups. He directed the budgetary largesse of road building to forge alliances with rural communities, framing his commercial objectives with the mantle of "progressive" and "revolutionary" ideals. He characterized the building of *caminos vecinales* as a means to "democratize" economic growth, incorporating into his administration's political platform of national unity many of the rural and working-class groups that his agricultural and labor policies excluded. By targeting patronage he strategically used road building to forge political alliances that ultimately benefited business allies by reducing the cost of shipping and increasing demand for motor travel. This approach was especially evident in how Alemán funded roads in northern Veracruz, appealing broadly to the region as a whole in campaign speeches, while federal agencies later approved construction plans that prioritized economically and politically important cities and towns over others.

Road-building policy was a tool for strategic trade-offs and deal making. On the one hand, left-wing national politicians used it to indicate willingness to work with elite private investors, while also reinforcing their commitments to state governments and rural non-elite groups. On the other hand, conservative politicians carried on with the language of revolutionary idealism as they used financial support for roads to co-opt amenable opponents in favor of commercial and pro-American objectives. Studying the impact of road policy thus uncovers the multilayered quality of national politics, highlighting clearly how power worked at different levels—and in distinct ways—in twentieth-century Mexico.

Exercising Power

The negotiations, conflicts, and compromises that occurred as part of the process of building new roads reaffirm Jeffrey Rubin's assertion that the Mexican state relied on "a distinct combination of bargaining, coercion, and alliances" to assert its power.[4] Road building did not occur as a top-down bureaucratic procedure directed entirely from Mexico City; it was activity mediated through the states. Federal officials coordinated with state governors and local road committees to formulate budgets, determine potential routes, recruit labor, and construct motorways. SCOP and its National Road Commission, and later the CNCV, bargained over the cost of this work, forging a cooperative model that shared power and formed political loyalties with state and local caciques. Governors remained the principal benefactors of this approach, experimenting with different kinds of organizational structures for road building and doling out patronage to curry regional support for new projects.

This study has shown that coercive policies were most evident in the interaction between state governments and local communities, where top-down and bottom-up political tensions marked regional interactions. Public officials sought to control how towns directed tax money collected for road building, while state engineers expropriated land and defined acceptable usage of motorways. In response ordinary citizens ignored rules, blocked roads, sued the government, or protested in other ways. Although federal officials threatened to cut funding when political gridlock arose, they had already ceded much of the decision-making power to state road boards.

The logistical and budgetary challenges of trying to build a national road network favored coalition building with state partners. From 1933 onward federal authorities used cost-sharing programs to pool financial resources and, in return for state-level contributions, allowed governors a greater say in the regional contours of national road policy. In the 1940s national and state leaders in the Partido de la Revolución Mexicana (Party of the Mexican Revolution) and later the Partido Revolucionario Institucional emphasized the language of modernity and progress in the guise primarily of economic modernization. This message appealed to regional conservatives long suspicious of the government's intentions for revolutionary social change in education, land reform, and anticlericalism.[5] By tying road building to pro-business economic modernization, public officials and party leaders fashioned it into a developmental program that reconciled the Mexican state with the country's powerful political Right.

Nevertheless local participation in the political process of roadwork reiterated the possibilities for communal agency. The handling of disputes highlighted the uneven power relationships tied to the negotiations behind roadwork, where state officials wielded significant bureaucratic authority, but ordinary citizens did not hesitate to fight back creatively. Rural organizations, labor unions, and drivers' cooperatives formed coalitions with state legislators and wealthy regional elites who could force road-building authorities to revise construction plans, adjust tax policy, or provide indemnities in the case of damaged or expropriated property. Local communities also played a critical role in forming road boards that helped state agencies to make sense of regional infrastructural needs, provide input for land surveys, and ultimately distribute resources to maintain and repair motor routes.

When local groups disagreed over national or state plans for a road, they exercised power to slow construction work, requiring political and legal intervention to sort out matters. They hired lawyers to lodge suits against road-building authorities in state and federal court and appealed directly to elected officials to intervene on their behalf. Local actors did not always speak with one voice politically, however; regional disagreements could generate competing factions that pursued different visions for a planned motorway. When this occurred state authorities were caught between fierce local rivalries, forcing public officials and the courts to work through a series of bitter recriminations to determine the final course of action. Although disagreements erupted among farmers, workers, industrialists, and business people, many still agreed generally with the official premise of new motorways. Even as they articulated the need for road building in different ways, many elite and nonelite Mexicans saw it as essential to the promotion of economic prosperity in the country.

Assessing the Legacy of Road Building in Mexico

In November 2014 I returned to Veracruz to visit friends and family. From Heriberto Jara International Airport, under renovation ahead of the Central American and Caribbean Games, which the state would host in a few weeks, I took a bus operated by Autobuses de Oriente to Xalapa. The highway that ran up the coast into the mountains was marked with signs announcing "Camino Real

1810–2010," commemorating Mexico's bicentennial. What struck me most about this highway, however, was a 30-meter portion, ten minutes from the airport. For years it had remained unpaved as construction crews slowly built a traffic bypass. In the meantime local men directed the flow of vehicles and asked for donations from the passing cars and pickup trucks. This pockmarked interlude on an otherwise fully paved road was as symbolic of the country's history of transport mobility as the official signage evoking thoughts of the royal road that Hernán Cortés had staked out centuries before. In many ways the highway between Xalapa and Veracruz was its own kind of monument to state formation in Mexico and a reminder that it remains an unfinished story.

In the years of economic and political turmoil that followed the Revolution, public officials had stoked a love affair for motorways that stayed with the nation for decades. When Venustiano Carranza called on the Mexican state to rebuild and expand the nation's road system, his administration faced a difficult task. Years of armed conflict and institutional neglect had left the nation's road infrastructure in terrible condition. Álvaro Obregón launched the first comprehensive program to build highways, but it was his successor, Calles, who established the national institutions that fulfilled Carranza's ambitions. Road building in Mexico was wrought with national significance that extended beyond purely economic conditions. It became a symbol and a metric of material progress and a means for national, state, and local actors to articulate whether the government was achieving the Revolution's promise to build a prosperous and equitable society.

This process was not without its contradictions. Road building brought deeper ties between the United States and Mexico, as well as greater private-sector investment. Ironically President Calles's decision in 1925 to establish a new road-building bureaucracy that pursued construction efforts as a nationalist endeavor exclusive of foreign financial commitments was gradually transformed into one of the critical spaces for Mexico and the United States to forge stronger binational ties. The building of new border highways created tangible connections between the two nations, which became important to diplomatic negotiations over oil expropriation and mutual aid in World War II. In the early postwar era, policymakers used road building to foster political consolidation under the aegis of the PRI, promoting market capitalism and friendship with United States.

In 1952, writing in Monterrey's *El Norte*, Ángel Lascuraín Osio, a well-known local engineer who contributed to a number of newspapers and magazines, reviewed thirty years of economic development in Mexico. He pointedly criticized Veracruz's governor Adalberto Tejeda and his support of agrarian activism that had called on campesinos to assert their rights as citizens and demand more control over the land. Lascuraín claimed that this strategy had diminished opportunities for rural economic development because it attacked wealthy landed interests. Instead he applauded Nuevo León's pro-business policies as the correct path for the nation.[6]

These criticisms reflected how the governments of Nuevo León and Veracruz had offered competing visions for economic and infrastructure development. Both states were economically important to the nation and served as key export corridors. Yet not only were they very different in terms of demographics, but public officials in Nuevo León and Veracruz pursued contrasting political philosophies that influenced road-building policy. In Veracruz the radical model that Governor Tejeda championed relied on campesinos to take an active role in road building, while in Nuevo León state leaders had long utilized public-private partnerships that used contractors to carry out the work. These differences narrowed over time, and by the 1940s Mexico had reached a consensus on road-building policy modeled on the practices that Monterrey elites had favored for over two decades.

Although national and state politicians articulated the benefits of economic growth in more artful, populist terms than did Lascuraín, the end goal was largely identical. The Mexican state's building of roads and other infrastructure was intended to support the needs of private enterprise and regional industrialization. By the early 1960s President Adolfo López Mateos could boast that federal officials had raised a record 300 million pesos in road bonds for new highways. His administration also launched a government-owned company to build more than 1,000 kilometers of toll roads in central Mexico. This venture later became an important means for greater private-sector involvement in construction efforts.[7]

Road-building programs continued to be a way for federal and state governments to forge strategic alliances with rural communities. Annual spending on federal and state highways and *caminos vecinales* reached 3.5 billion pesos in 1973, and roughly one-third of those funds were allocated to projects in the countryside. President Luis Echevarría said, "We have continued road construction with direct participation from rural workers. The advantages this labor offers is suitable to our needs, becoming a driver of rural development. . . . These works have benefited more than 1.1 million compatriots."[8] Between 1988 and 1994 the government under Carlos Salinas de Gortari built more than 10,000 kilometers of rural highways and *caminos vecinales* as part of his Programa Nacional de Solidaridad (National Solidarity Program).[9] Even as he privatized toll road operations and doled out lucrative road-building contracts to construction companies, he wrapped these activities in the language of rural improvement. Not unlike Miguel Alemán, he framed road building as a populist good that benefited workers and the poor, while aggressively expanding corporate power in Mexico.

After the 2000 presidential election, when the PRI's longtime rival, the Partido de Acción Nacional (National Action Party), finally gained power, it carried on with the political legacy of road building as a measure of modernization. President Vicente Fox channeled hundreds of millions of pesos into road building via his program for basic infrastructure needs. Likewise Felipe Calderon, under the banner of building "a better and more connected nation," touted the construction and renovation of more than 23,000 kilometers of roads, a project the federal government undertook with the Comisión de Pueblos Indígenas to combat endemic rural poverty.[10] The allure of seeing new highways as monuments to political success was too difficult for many leaders to resist.

But road building alone could not address the economic and political problems facing much of the country. Neoliberal policies and free trade industrialized the northern border, while economic crises in the 1980s and 1990s unsettled the nation's working and middle classes. Millions of people left their homes in central and southern Mexico, often emigrating long distances by bus for faraway jobs in other parts of the country or the United States.[11] Economic hardship, compounded by years of armed civil strife following Calderon's war on the drug cartels in 2008, added another layer of complexity to the story of Mexico's roads and highways. In this conflict they became sites of intense violence and social protest as different groups created roadblocks that shut down major thoroughfares, blocked regional motor traffic, and captured the attention of national and international media outlets.[12]

As the government points to road building as symbolic of economic progress, other people will dispute this narrative, converting roads and streets into spectacles of gridlock, burned buses, and political struggle. It is indicative of economic and social tensions rooted in the very idea of Mexico itself, dating not just to the causes of the Revolution but also to national independence and the legacy of colonialism. Road building will remain deeply contested, as it is a political and technical project that reaches into the heart of communities and regions, weaving them into a national fabric marked by deep social divisions and economic inequality.

Looking back on its history, I find that road building in Mexico certainly achieved the first half of presidential promises made since 1917. The roads connected regional production centers to urban industrialization, gradually becoming more closely integrated with transnational markets over the course of three decades. It is much less clear whether they distributed this prosperity in an equitable fashion. On the one hand, many local groups lobbied for and built motor roads that extended the reach of their communities, improving regional mobility and providing access to neighboring cities and ports. On the other hand, logistical and budgetary limitations, as well as the impact of corruption, left many rural parts of the country underserved by regional transportation infrastructure for years.

For many people, new highways and roads were the sine qua non of material improvement. They believed in the potential for motorways to provide dramatic change in regional mobility and material conditions. Road building was a critical factor in the country's social, political, and economic future after the Revolution. Throughout this history wide-ranging groups of citizens worked with and contested construction efforts that ultimately helped to define fundamental aspects of the state and society in Mexico in the twentieth century. In many respects this project continues.

APPENDIX A

Comparing the Real Cost of Federal and State Spending on Roads

The kinds of roads built in a given year offer insight into the priorities that political leaders and bureaucratic officials applied to construction efforts. As shown in table 1, between 1925 and 1933 federal and state governments nationwide built dirt, macadam, and asphalt roads at almost an equal rate. Only after 1933 do the data show an acceleration in the rate of construction for routes with macadam and asphalt surfaces. In fact into the 1940s, as the number of dirt and macadam roads dropped, asphalt roads continued to increase at a steady pace. This relationship corresponds with archival evidence indicating greater demand among local communities and businesses for the construction of better quality roads to support bus service and resist poor weather conditions. The data show clearly that the state responded to these demands; by 1952 the country counted more than seven times as many asphalt concrete as dirt roads.

Table 1. Federal and state road-building surface type
(in kilometers, accumulated)

	Dirt	Macadam	Asphalt	Total
1925–28	209	245	241	695
1929	353	298	289	940
1930	629	256	541	1,426
1931	683	377	620	1,680
1932	802	367	645	1,814

1933	1,601	793	683	3,077
1934	1,786	1,291	1,183	4,260
1935	1,760	1,918	1,559	5,237
1936	1,891	2,406	2,007	6,304
1937	1,831	3,363	2,316	7,510
1938	2,035	3,424	3,004	8,463
1939	1,912	3,441	3,755	9,108
1940	1,643	3,505	4,781	9,929
1941	2,249	4,131	5,420	11,800
1942	2,250	5,194	6,082	13,526
1943	2,418	5,918	9,610	15,246
1944	2,336	6,375	7,683	16,394
1945	2,399	6,842	8,163	17,404
1946	2,663	7,267	8,614	18,544
1947	2,509	7,722	9,071	19,302
1948	2590	6,775	10,562	19,927
1949	2,453	5,927	12,059	20,439
1950	1,865	5,972	13,585	21,422
1951	2,034	5,857	14,980	22,871
1952	2,039	5,905	15,981	23,925

Source: INEGI, *Anuario estadístico de los Estados Unidos Mexicanos*, 1930, 1939, 1943–45, 1955–56, 1957, 1958, http://www3.inegi.org.mx/sistemas/biblioteca/ficha.aspx?upc=702825087340 (accessed January 25, 2017); Nacional Financiera, *50 años de revolución mexicana en cifras*.

As changes in construction policy favored better quality roads, official spending increased to pay for it. In table 2 the federal budget shows marked increases in spending on new roads over a thirty-year period. Three exceptions occur: between 1932–34, due to the impact of the Great Depression and the creation of the Program for Cooperation on Roads, which reduced the federal government's commitment to funding new roads; 1939–40, likely due to expropriation of the oil sector and the subsequent U.S. embargo that burdened the Mexican economy; 1944. It is unclear why spending dropped in 1944, although it may have been due to material and gasoline shortages that had a structural impact on national economic development during World War II. Otherwise there is continued growth in the federal spending on roads.

Table 2. Federal spending on roads (in pesos)

	Construction	Maintenance	Total
1925–28	31,381,436	no data	32,581,436
1929	8,209,940	no data	8,209,940
1930	12,042,342	1,301,050	13,343,392
1931	12,975,028	889,000	13,864,028
1932	5,837,658	726,619	6,564,277
1933	8,403,391	1,404,309	7,807,700
1934	8,213,730	1,501,943	9,715,673
1935	15,364,510	2,002,008	17,366,518
1936	27,508,693	3,396,591	30,905,284
1937	28,966,711	3,726,444	32,693,155
1938	30,617,062	2,845,784	34,050,038
1939	24,091,450	3,432,976	29,955,391
1940	23,661,159	5,863,941	33,928,423
1941	41,821,910	10,267,264	54,161,934
1942	91,862,289	12,340,024	109,386,488
1943	111,562,511	17,524,199	127,048,512
1944	95,408,284	15,486,001	116,943,417
1945	100,552,874	21,535,133	124,429,087
1946	12,732,225	23,876,213	153,967,500
1947	126,518,259	26,645,275	156,138,707
1948	183,396,678	29,620,448	222,240,828
1949	210,792,556	38,844,150	250,114,829
1950	211,959,569	39,322,273	256,402,508
1951	218,348,035	44,442,939	279,178,818
1952	377,686,418	60,830,783	377,686,418

Although the normal data on construction and spending indicate robust growth, when the cost is calculated per kilometer and adjusted for inflation, a much more complicated view emerges.[1] The second column of table 3 shows the annual spending on road building per kilometer according to government data recorded by INEGI. By adjusting for inflation (fourth column, table 3), we can see more clearly the real amount of spending on road construction. Even as spending increased in moments of pronounced fluctuations in the cost per kilometer, overall the expenditures did not change as dramatically as the official, nominal data suggested. For example, in 1929, prior to the cost-sharing Program for Cooperation on Roads, the federal government actually spent more on construction than it did in 1952. Likewise between 1932 and 1933 deflationary pressure on the peso actually made road building more expensive even as federal officials drastically cut spending and pushed costs to state governments. When the impact of inflation is factored into the analysis, the real amounts spent on road building during the Ávila Camacho and Alemán *sexenios* appear less impressive than what the contemporary political rhetoric touted.

Table 3. Average federal spending on roads

Cost per kilometer (in pesos), 1929=100

	Nominal Cost	Wholesale Price Index	Inflation-adjusted real cost in 1929 pesos	Cost variation per year
1925–28	45,153.15	100.30	45,018.09	
1929	33,509.26	100.00	33,509.96	-11,508.13
1930	24,778.48	100.20	24,729.02	-8,780.94
1931	51,082.79	90.00	56,759.65	32,029.63
1932	43,564.61	82.60	52,741.77	-4,016.99
1933	5,069.98	88.30	5,741.77	-46,999.89
1934	6,943.14	91.20	7,613.09	1,871.32
1935	15,726.21	90.60	17,357.85	9,744.76
1936	25,781.34	117.77	21,891.19	4,533.34
1937	24,018.83	138.96	17,284.40	-4,606.79
1938	32,127.03	147.35	21,803.06	4,518.66
1939	37,351.09	148.23	25,197.38	3,394.32
1940	28,819.93	151.99	18,962.16	-6,235.25
1941	22,352.70	162.14	13,785.94	-5,176.19
1942	53,222.65	178.70	29,783.64	15,997.70
1943	64,861.93	216.11	30,312.72	229.08
1944	83,108.26	264.57	31,412.63	1,399.92
1945	99,557.30	294.59	33,795.02	2,382.39
1946	111,686.16	339.40	32,906.56	-888.47
1947	166,910.63	350.55	47,613.68	14,707.12
1948	293,434.68	385.65	76,088.10	28,474.43

1949	411,704.21	422.52	97,440.97	21,352.86
1950	215,625.20	461.82	46,691.31	-50,749.66
1951	150,688.78	572.63	26,315.35	-20,375.96
1952	358,336.26	593.49	60,378.03	34,062.68

Source: INEGI, *Anuario estadístico de los Estados Unidos Mexicanos*, http://www3.inegi.org.mx /sistemas/biblioteca/ficha.aspx?upc=702825087340 (accessed January 25, 2017).

If this were the end of the story it would explain why local communities had complained about a lack of funds. Inflation had eaten into the amount of money the government had available to dedicate to construction efforts. Yet that is not entirely the case. For example, as shown in table 3, even when adjusted for inflation, there are marked increases in the real amount of money spent per kilometer of road in 1942, 1948, 1949, and 1952. Although the government had shifted to the construction of more asphalt roads, these sharp spikes in cost are not entirely explained by this change in policy.

Although it is difficult to fully ascertain why these increases occurred we can speculate on two important correlations. By 1942 Maximino Ávila Camacho, in control of SCOP, had done away with more than a decade of incremental expenditures and pursued a more spendthrift course thanks to the injection of millions of dollars in investment during World War II. In 1948, with the creation of the Alemán's CNCV, the government gave large sums of money to private contractors linked to friends of the president. The fluctuations in the per-kilometer cost may indicate a politically driven motive as officials coped with what may have amounted to a major giveaway to government cronies. Clearly the historical record has shown both of these presidential terms were marked by acute levels of corruption. Analysis of the data appears to support this assessment.

In this context the data behind road-building efforts reveal a complicated set of factors. Public officials showed that the leviathan was willing to listen when they adapted policy priorities to favor popular and private-sector demands for better quality roads. From 1935, as the road-building bureaucracy became more professionalized, until 1941, when Maximino Ávila Camacho took over SCOP, a period of conservative but consistent amounts of spending funded construction efforts. By the mid-1940s, however, budgets and per-kilometer costs fluctuated considerably. The decision to increase the number of asphalt roads certainly caused costs to rise, while the growing dependence on foreign and private investment also likely played a role in this issue. Nevertheless it appears that government corruption and cronyism may have been responsible for some of the dramatic spikes in per-kilometer costs during this time, given what we know of the political atmosphere of the 1940s and early 1950s.

Table 4. Index of wholesale prices, Mexico City, 1928–1941 (in pesos)

1929=100

1928	100.30
1929	100.00
1930	100.20
1931	90.00
1932	82.60
1933	88.30
1934	91.20
1935	90.60
1936	95.50
1937	112.30
1938	115.80
1939	118.50
1940	120.20
1941	127.20

Source: INEGI, *Anuario estadístico de los Estados Unidos Mexicanos*, http://www3.inegi.org.mx /sistemas/biblioteca/ficha.aspx?upc=702825087340 (accessed January 25, 2017).

APPENDIX A

Table 5. Index of wholesale prices, Mexico City, 1936–1952 (in pesos)

1935=100

1936	106.70
1937	125.90
1938	133.50
1939	134.30
1940	137.70
1941	146.90
1942	161.90
1943	195.80
1944	239.70
1945	266.90
1946	307.50
1947	317.60
1948	349.40
1949	382.80
1950	418.40
1951	518.80
1952	537.70

Source: INEGI, *Anuario estadístico de los Estados Unidos Mexicanos*, http://www3.inegi.org.mx /sistemas/biblioteca/ficha.aspx?upc=702825087340 (accessed January 25, 2017).

Table 6. Adjusted index of wholesale prices, Mexico City, 1928–1952 (in pesos)

1929=100

1928	100.30
1929	100.00
1930	100.20
1931	90.00
1932	82.60
1933	88.30
1934	91.20
1935	90.60
1936	117.77
1937	138.96
1938	147.35
1939	148.23
1940	151.99
1941	162.14
1942	178.70
1943	216.11
1944	264.57
1945	294.59
1946	339.40
1947	350.55
1948	385.65
1949	422.52
1950	461.81
1951	857.63
1952	593.49

Note: This index, which is used to adjust for inflation, was created by combining historical data from INEGI, *Anuario estadístico de los Estados Unidos Mexicanos*, http://www3.inegi.org.mx/sistemas/biblioteca/ficha.aspx?upc=702825087340 (accessed January 25, 2017).

Comparing the Budgets for Program for Cooperation on Roads
and the Comisión Nacional de Caminos Vecinales

Scop's Program for Cooperation on Roads and the Comisión Nacional de Caminos Vecinales were important mechanisms for federal and state coordination. In different ways both entities directed millions of pesos in spending on road construction to local communities. Their composition was distinct and their results varied over the years they existed. Administratively the Program for Cooperation on Roads carried out its activities within the established road-building bureaucracy in scop and distributed funds to state-level agencies and the federal jlcs. In contrast, the cncv largely operated as a separate entity from scop, under greater influence by the president and the executive committee, which directed its activities. Although the Program for Cooperation on Roads was susceptible to corruption and political interference, it remained a more conservative venture, restrained by bureaucratic protocols. The cncv, given its organizational structure, may have been even more vulnerable to the individual whims of the people who controlled it.

Although this distinction created opportunities for politically connected contractors, during the years covered in this study the Program for Cooperation on Roads retained a significantly larger amount of spending. As table 7 shows, federal and state expenditures vastly outweighed the amounts spent by the cncv. Not only did the former spend more, but this amount increased over time even as the latter was established and began to carry out roadwork after 1948. Unfortunately it is difficult to ascertain annual changes in CNCV spending as INEGI recorded expenditures for the period 1948–52 as a single figure. Nevertheless it is clear that although the CNCV enjoyed close ties to the president and benefited some of his personal associates, it operated at a much smaller scale than the program in SCOP. In this context the CNCV appears to have possibly existed as a mechanism for the president to reward friends and allies, while not disrupting the overall work of the existing road-building bureaucracy.

Table 7. Federal and state spending on Program for Cooperation on Roads (in pesos)

1929=100

	Nominal Federal Spending	Nominal State Spending	Wholesale Price Index	Inflation-Adjusted Real Federal Spending	Inflation-Adjusted Real State Spending
1933	3,604,223.00	2,267,872.00	88.30	4,081,792.75	2,568,371.46
1934	5,055,889.00	3,729,823.00	91.20	5,543,737.94	4,089,718.20
1935	6,237,131.00	4,701,753.00	90.60	6,884,250.55	5,189,572.85
1936	8,760,718.00	5,832,954.00	117.77	7,438.810.22	4,952,817.55
1937	15,438,534.00	9,962,484.00	138.96	11,109,858.46	7,169,190.23
1938	11,979,526.00	10,052,057.00	147.35	8,129,925.51	6,821,845.42
1939	10,071,668.00	9,572,810.00	148.23	6,794,438.73	6,457,904.59
1940	14,919,000.00	8,131,453.00	151.99	9,815,986.93	5,350,106.33
1941	22,318,112.00	9,573.804.00	162.14	13,764,608.22	5,904,606.14
1942	24,788,820.00	20,371,673.00	178.70	13,871,940.04	11,400,083.84
1943	26,313,818.00	15,850,120.00	216.11	12,175,852.46	7,334,120.90
1944	26,324,159.00	18,735,252.00	264.57	9,949,807.28	7,081,409.39
1945	25,445,902.00	17,795,414.00	294.59	8,637,687.23	6,040,706.29
1946	31,952,961.00	18,233,483.00	339.40	10,298,335.83	5,372,206.70
1947	47,418,380.00	18,781,266.00	350.55	13,526,779.69	5,357,628.15
1948	51,496,680.00	22,285,032.00	385.65	13,353,174.61	5,778,545.79
1949	39,934,491.00	23,842,979.00	422.52	9,451,580.16	5,643,087.51
1950	53,617,145.00	24,755,000.00	461.81	11,610,213.52	5,360,427.82
1951	85,836,347.00	42,718,569.00	572.63	14,989,924.90	7,460,104.76
1952	104,948,699.00	38,094,580.00	593.49	17,683,377.59	6,418,763.15

Table 8. Comisión Nacional de Caminos Vecinales spending

	Nominal CNCV Spending	Nominal State Spending	Wholesale Price Index	Inflation-Adjusted Real CNCV Spending	Inflation-Adjusted Real State Spending
1948–52	48,558,255.00	58,388,759.00	461.18	10,514,769.06	12,643,459.21

Minimum Wages in Nuevo León and Veracruz for Road Workers

The graph shows notable differences in the minimum wages of roads in Nuevo León and Veracruz. This distinction is related to the organizational structure of the road-building bureaucracy in these states. In Nuevo León one agency worked with a single union to carry out projects and negotiate pay scales. This unified arrangement created greater stability and limited steep rises in worker pay. In contrast, political conditions in Veracruz prevented state officials from centralizing road-building efforts. Instead multiple agencies, regional committees, and private companies carried out this work. By failing to exercise greater control over the process, state officials essentially allowed these organizations to compete with one another for resources to build motorways. As such it is likely that market competition among organizations was responsible for a higher minimum wage for road workers in Veracruz.

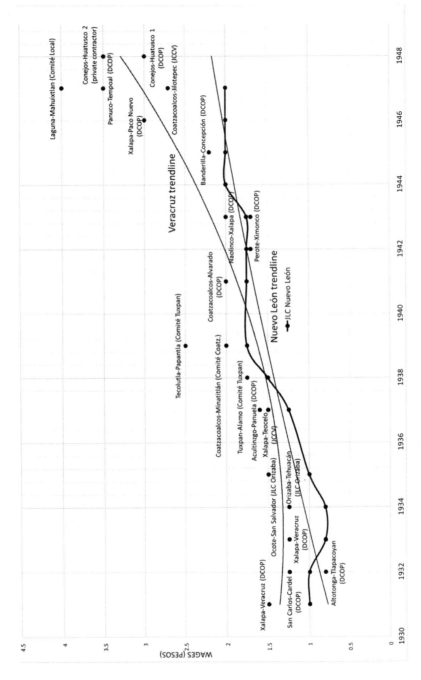

Road Worker (peon) Daily Minimum Wage by Agency and Project, 1931–1948. Archivo General del Estado de Veracruz, Archivo General del Gobierno de Nuevo León.

NOTES

Introduction

1. Venustiano Carranza, second report to the government, 1 September 1918, 106; third report to the government, 1 September 1919, 199, Informes Presidenciales (IP), Cámara de Diputados (CD), Dirección de Servicios de Investigación y Análisis (DSI), México, DF (2006). All translations are mine.

2. Secretaría de Obras Públicas, *Caminos y Desarrollo*, 24–26; "Construcción y reconstrucción de caminos," *El Universal*, 8 February 1921; John Stilgoe et al., "Roads, Highways, and Ecosystems," *Encyclopedia of Earth*, https://nationalhumanitiescenter.org/tserve/nattrans/ntuseland/essays/roads.htm (accessed 16 January 2017). For figures on car ownership in the United States, see Drowne and Huber, *The 1920s*, 244.

3. Waters, "Remapping Identities," 221–42.

4. Knight, "Introduction," 8–33.

5. Scott, *Seeing Like a State*; Weber, *Peasants into Frenchmen*.

6. Rubin, "Decentering the Regime"; Vaughan, *Cultural Politics in Revolution*.

7. Joseph and Nugent, *Everyday Forms of State Formation*.

8. Knight, "Weapons and Arches in the Mexican Revolutionary Landscape," 24–66.

9. Joseph and Nugent, "Popular Culture and State Formation in Revolutionary Mexico," 19.

10. Joseph and Nugent, "Popular Culture and State Formation in Revolutionary Mexico," 12–22.

11. John Welsh, "Good Roads Lead into the Mexican Republic," *New York Times*, 24 February 1929; telegrams and letters on Calhoun Highway Association in Fondos Presidenciales (FP) Obregón-Calles, 1921–28, volumen 27, expedientes 104-C-60, 104-C-139, Archivo General de la Nación (AGN), Mexico City; Miguel W. Guerrero, chief litigator for Huasteca Petroleum Company (HPC), complaint against government of Veracruz, 11 October 1930; state response to HPC, 18 October 1930; order to Attorney General of Veracruz, 17 November 1930, Fondo Legal (FL), 1930–35, box 86, Archivo General del Estado de Veracruz (AGEV), Xalapa.

12. Waters, "Remapping the Nation."

13. Freeman, "'La carrera de la muerte'"; Freeman, "'Los hijos de Ford.'"

14. Berger and Grant Wood, *Holiday in Mexico*; Freeman, "Driving Pan-Americanism."

15. Berger and Grant Wood, "Introduction," 1–9.

16. My recent publications, which have helped to develop and conceptualize this book, are "Revolutionary Paths" and "Routes of Conflict."

17. In *Mexico in the 1940s*, Stephen Niblo identifies six major categories: direct looting, abuse of authority, kickbacks, hidden deals, payments for services rendered, and inside information.

18. Niblo, *Mexico in the 1940s*, 253–310.

19. Seminario Judicial, Amparos (1930–39), Archivo Histórico de la Suprema Corte de la Justicia de la Nación (AHSCJN), Mexico City.

20. For more on bus drivers as new industrial workers, see Freeman, "'Los hijos de Ford.'"

21. Anita Brenner, "Old Mexico Changed by a New Highway," *New York Times Magazine*, 18 August 1935; "Transporte automotriz la base del progreso," *El Porvenir*, 5 September 1942.

22. Charles Clifton, "The World Needs Motor Transportation," *New York Times*, 4 January 1925.

23. Booth, "Turismo, Panamericanismo e Ingeniería Civil"; Ballent, "Ingeniería y Estado"; Wolfe, *Autos and Progress*.

24. Kim, "Destiny of the West."

25. Luis Mayer, "Los modernos caminos crean nuevos valores," *Excélsior*, 30 August 1925.

26. "Las buenas carreteras harán aumentar el automovilismo," *Excélsior*, 16 August 1926.

27. Coatsworth, *Growth against Development*, 7–11; Coatsworth, "Railroads, Landholding, and Agrarian Protest," 56; Kuntz-Ficker, *Empresa extranjera y mercado interno*, 355–59; Matthews, *The Civilizing Machine*, 23–54.

28. Mora-Torres, *The Making of the Mexican Border*, 65–66, 87–89.

29. "Como harán la construcción de carreteras," *Excélsior*, 12 August 1925.

30. Rubén Bauchez, letter to Cerdán, 28 July 1943, box 1060, exp. 317/70 Huiloapan-Nogales, AGEV; for full quote see chapter 5.

31. Van Hoy, *A Social History of Railroads in Mexico*, 8–19.

32. Junta Local de Caminos de Nuevo León (JLCNL), Personnel Report, 1 July 1940, box 26; Manuel Saldaña, letter to Pablo Domínguez from the Sindicato de Empleados y Obreros Constructores de Caminos del Estado de Nuevo León, 3 August 1942, box 34, JLCNL, Secretaría de Comunicaciones y Obras Públicas, Archivo General del Gobierno de Nuevo León (AGGNL), Monterrey.

33. Freed, "Network of (Colonial) Power," 203–23.

34. For more information on this contrast in pay rates between Nuevo León and Veracruz, see my graph "Road worker (peon) daily minimum wage by agency and project, 1931–1948," in appendix C.

35. Ben Fulwider in "Driving the Nation" first noted that Cárdenas pursued relatively conservative, pro-business policies for road building during his *sexenio*.

1. "A Good Road . . . Brings Life"

The chapter title is a translation from the Spanish: "Una amplia vía a través de nuestros hermosos y ricos bosques, será la atería que lleve la vida a todos los pueblos por donde pase." Carlos Barrios, "Carretera Zaragoza-Tecolutla: Hermas Obras," 6 November 1921, FP Obregón-Calles, box 335, AGN.

1. "Excursión de Veracruz a Jalapa en Autos Overland," *El Universal*, 16 July 1922. For the history of the road to Xalapa in the colonial period and the nineteenth century, see Moore, *Forty Miles from the Sea*.

2. "Comentarios del día: La construcción de carreteras," *El Dictamen*, 20 April 1922.

3. "Excursión automovilística de México al Puerto de Veracruz," *Excélsior*, 26 March 1921.

4. "Pronto surgirá una red maravillosa de caminos carreteros," *Excélsior*, 4 September 1917; "Una gran carretera para dar salida al petróleo de la región de Tuxpan," *El Universal*, 26 March 1919; "El automovilismo y los caminos carreteros," *Excélsior*, 26 October 1919.

5. "Informaciones de la República: Jalisco," *Excélsior*, 30 January 1920; "Comentarios del día," *El Dictamen*; "Excursión de Veracruz a Jalapa en Autos Overland," *El Universal*; "Urge una ley de caminos," *Excélsior*, 6 April 1920; Freeman, "'Los hijos de Ford,'" 225.

6. The Huasteca Veracruzana is a region of northern Veracruz that includes the port of Tuxpan, which was a major site of oil production in the early twentieth century. Alvaro Obregón, first state of the union address, 1 September 1921, IP, CD, DSI, Mexico, DF (2006), 45.

7. Aboites and Loyo, "La construcción del nuevo Estado," 595–615; Luis Anaya Merchant, "Guerra, automóviles y carreteras. La influencia norteamericana y el Mercado automotriz mexicano en la 'reconstrucción' posrevolucionaria," *Boletín N. 73* of the Fideicomiso Archivos Plutarco Elías Calles y Fernando Torreblanca (May–August 2013).

8. Obregón, second state of the union address, 1922, IP, CD, DSI, Mexico, DF (2006), 97.

9. "El camino meridiano," *El Universal*, 3 June 1922.

10. Chamber of Commerce of Laredo, letter to Cámara Nacional de Comercio de Monterrey, 8 March 1921; Departamento de Justicia, Instrucción Pública y Fomento, letter to Secretaría de Gobierno, 13 April 1921, box 1, Comisión Nacional de Caminos Vecinales (CNCV), 1921–59, AGGNL, Monterrey, Nuevo León.

11. Cámara Nacional de Monterrey, letter to Nuevo León Secretaría de Gobierno, 18 April 1921; SCOP, letter to Governor Juan M. García, 5 April 1921, box 1, CNCV, 1921–59, AGGNL.

12. *Boletín de la Junta de Mejoras Materiales*, No. 1, 23 August 1926, Secretaría de Gobernación (SEGOB), Fomento, 1925–30, Mejoras y Obras Públicas, box 212, AGEV.

13. Calixto González, letter dated 14 January 1931, Fomento, 1926–33, Comunicaciones y Obras Públicas (COP), box 20, Informes y Caminos (Caminos), AGEV.

14. Freeman, "'Los hijos de Ford,'" 225; "Historia de *El Porvenir*: 94 años de Hablar con la Verdad," editorial, *El Porvenir*, 27 June 20152013,https://web.archive.org/web/20150627181652/http://www.elporvenir.com.mx/index.php?option=com_content&view=article&id=96&Itemid=398 (accessed 14 January 2017); Pussetto, Vázquez, and Esparza, "Análisis de la ideología empresarial regiomontana," 11–13.

15. "Necesitamos carreteras," *El Porvenir*, 4 August 1926: "Nuestro país no ha podido apreciar, en toda magnitud, al beneficio de las carreteras, porque no ha podido disfrutar de estas vías de comunicación. En los paises modernos los caminos reciben desde hace mucho y están recibiendo impulso superior a toda otra mejora de interés colectivo, porque la riqueza nacional, la economía de los pueblos se ha sentido estimulado de modo decisivo."

16. "Necesitamos carreteras," *El Porvenir*.

17. Luis Mayer, "Los modernos caminos crean nuevos valores," *Excélsior*, 30 August 1925.

18. Vaughan, *Cultural Politics in Revolution*, 4–8.

19. Barrios, "Carretera Zaragoza-Tecolutla."

20. Barrios, "Carretera Zaragoza-Tecolutla," italics added: "La idea hermosa que han tenido algunas personas amantes del progreso es elevada y es grande. . . . Una amplia via a través de nuetros hermosos y ricos bosques, será la arteria que lleve la vida a todos los pueblos por donde pase. . . . La carretera Zaragoza-Tecolutla será hecha por el pueblo y para su beneficio propio . . . la haremos todos los hombres conscientes y amantes de nuestro terruño, y terminada que sea, cada pueblo por donde pase, cobrará el tráfico respectivo, dedicando ese fondo a la conservación de la propia via y al sostenimineto de la Educación Pública."

21. Efraín F. Bonilla, "Una muestra de verdadero progreso es la carretera que se construye de Zaragoza a Tecolutla," 15 November 1921, FP Obregón-Calles, box 335, AGN.

22. Coatsworth, "Railroads, Landholding, and Agrarian Protest," 49–56, 67–71.

23. O'Flaherty, *Highways*, 227–30; A. Hernández Bravo, "La pavimentación moderna," *El Universal*, 30 April 1922.

24. Obregón, IP 97.

25. "Lamont in Mexico for Bondholders," *New York Times*, 12 October 1921; "Obregón Sentiment Grows: Latin American Would Not Uphold Coercion on Treaty," *New York Times*, 25 January 1922.

26. "Obregón Signs Mexican Debt Agreement Negotiated by Lamont and de la Huerta," *New York Times*, 8 August 1922.

27. "Trabajo de construcción de carreteras que se suspenden," *Excélsior*, 19 April 1922.

28. "Una política que fracasó por falta de dinero," *El Universal*, 28 April 1922.

29. Municipal Council of Jonotla, Puebla, letter to Obregón, 24 June 1922; municipal council of Cuetzalan, Puebla, letter to Obregón, 22 September 1922; Obregón, multiple form letters addressed to leaders of Zaragoza-Tecolutla road, 19 October 1922, FP Obregón-Calles, box 335, AGN.

30. "Notas de Monterrey," *El Univerisal*, 4 April 1922.

31. Ramiro Támez, governor of Nuevo León, letter to SCOP General Secretary, 23 May 1922; Cámara Nacional de Comercio, Industria y Minería del Estado de Nuevo León, letter to Governor Támez, 24 May 1922; SCOP, letter to Támez, 20 June 1922; SCOP, letter to JLCNL, 14 August 1922; Támez, letter to Plutarco Elías Calles, secretario de gobernación, 2 March 1923, box 1, CNCV, 1921–59, AGGNL.

32. "La acción directa para una obra de general interés," *El Dictamen*, 20 September 1922.

33. "Una política que fracasó por falta de dinero," *El Universal*.

34. Fowler Salamini, *Agrarian Radicalism in Veracruz*, 25–40, 46–47, 88–107, 108–17.

35. "El gobernador, en su informe, se ocupa extensamente del desarrollo de los sindicatos y de la ayuda que les presta," *El Dictamen*, 17 September 1922; "Excursión de Veracruz a Jalapa en Autos Overland," *El Universal*.

36. Santiago, *The Ecology of Oil*, 102–13; Hacienda Chichilmantla, box 1686, exp. 47486, Veracruz, Archivo Histórico de PEMEX (AHP), Mexico City.

37. Lawyers for Huasteca Petroleum make reference to the 1842 law. See Third District Judge, 5 January 1931, FL, 1930–35, box 86, AGEV. The language of the 1842 statute is found in *Decretos y Ordenes de Interés Común que Dicto El Gobierno Provisional (Decretos)*, Tomo II, July 1842–June 1843, Mexico (J. M. Lara), 81–83. Harvard Law Library, https://books.google.com.mx/books?id=2BRTAAAAYAAJ&pg=PA81&lpg=PA81 &dq=Los+caminos+que+solo+vayan+a+las+haciendas+y+ranchos+se+consideran +privados&source=bl&ots=zqPbehT_nv&sig=hQg8yT6UIbBoWxSuzSVly-_Sy8s&hl =en&sa=X&redir_esc=y#v=onepage&q=Los%20caminos%20que%20solo%20vayan %20a%20las%20haciendas%20y%20ranchos%20se%20consideran%20privados&f= false (accessed 14 January 2017).

38. *Decretos*, 81–83: "Los caminos que solo vayan a las haciendas y ranchos se consideran privados."

39. Complaint against the HPC, 24 April 1922; Compañía Petrolera el Agwi, letter to the governor, 24 April 1922, box 498, AGEV.

40. Fowler Salamini, *Agrarian Radicalism in Veracruz*, xiv–xv, 25–29.

41. Hart, *Empire and Revolution*, 272–77, 342–49.

42. Box 2144, file 57952 (multiple years), AHP, includes newspaper articles, public correspondence, and court documents; "Un gran oleoducto del Puerto de Tampico a México," *El Universal*, 19 February 1920; "A las cias: Petroleras y al públicos en general," *Excélsior*, 3 September 1921; "Las guardias blancas que sostiene 'la Huasteca Petroleum Co.' son un peligro," *El Demócrata*, 29 January 1925; "Proyecto para substituir las guardias blancas con las tropas dadas de Baja," *Excélsior*, 1 February 1925; "Los vecinos de Juan Felipe Elevan una queja ante el Secretario de Industria, Sr. Morones," *El Demócrata*, 1 February1925.

43. Complaint against HPC, 24 April 1922.

44. Joaquín Cobos, letter to Reyes, municipal president of Tepezintla, against HPC, 26 April 1922, box 498, AGEV: "Por asuntos relacionados con nuestros intereses particulares y de familia, hemos tropezado con dificultades, en virtud de que la Compañía 'La Huasteca,' propietaria de los terrenos de Cerro Azul de esta misma jurisdicción ha colocado puertas en el camino . . . impidiendo el paso a toda persona, siempre que no presente el pase correspondiente del señor Ventura Calderón, encargado del Departamento de Terrenos de la expresada compañía, siendo así que el camino de referencia ha sido reconocido por todos los habitantes de los pueblos cercanos como nacional y no como camino privado según lo manifiesta la Compañía [sic]. Y siendo este procedimiento ilegal e injusto, por contravenir la sansión [sic] del artículo 11 de la Constitución General de los Estados Unidos Mexicanos . . . toda vez que en su espíritu, concede a todo hombre la garantía de entrar, salir y viajar por el territorio de la República sin necesidad de carta de seguridad, pasaporte, salvo conducto u otro requisito semejante."

45. Joaquin Cobos, letter to the governor of Veracruz against HPC, 27 April 1922;

HPC, telegram to the governor of Veracruz, 26 April 1922; official response from the governor, 17 May 1922, box 498, AGEV; Huasteca, box 2144, exp. 57952, Veracruz, AHP; Juan Felipe, box 2145, exp. 57966, Veracruz, AHP.

46. State government of Veracruz, letter to the federal government, 19 May 1922, box 498, AGEV.

47. Mexico-based subsidiary of the British-owned Atlantic Gulf and West Indies Petroleum Company, today a part of ExxonMobil, online: http://www.exxonmobil.co .uk/UK-English/files/Fawley_2011.pdf (accessed 22 January 2015).

48. William Green, writ of *amparo* against government of Veracruz, 13 September 1922; District Court of Tuxpan and Supreme Court of Mexico, court summary, 1922 and 1938 (multiple dates), box 498, AGEV.

49. Cobos, letter to the governor of Veracruz against HPC, 27 April 1922; Green, letter to Secretario de Gobierno de Veracruz, 26 April 1922, box 498, AGEV.

50. Green, writ of *amparo* against government of Veracruz, 13 September 1922; District Court of Tuxpan and Supreme Court of Mexico, court summary, 1922 and 1938 (multiple dates), box 498, AGEV.

51. Juan Felipe, box 2145, exp. 57966, Veracruz, AHP.

52. "General Calles Assumes Command," *New York Times*, 30 November 1924; "Como harán la construcción de las carreteras," *Excélsior*, 12 August 1925; Plutarco Elías Calles, state of the union address, 1 September 1925, IP, CD, DSI (Mexico City), 73.

53. "Mexico Begins Road Building," *Los Angeles Times*, 10 May 1925; "Como harán la construcción de las carreteras," *Excélsior*; Plutarco Elías Calles, state of the union address, 1 September 1925, IP, CD, DSI (Mexico City), 73; "Mexico May Get Loan to Build Roads," *Los Angeles Times*, 20 May 1925; "Como harán la construcción," *Excélsior*.

54. Manuel Gómez Morín, cited in Jean Meyer's "Revolution and Reconstruction," 218–19, cited in Gauss, *Made in Mexico*, 41. In her book Gauss includes only the same brief quote. She does not mention the date Gómez Morín made this statement.

55. "Como harán la construcción de las carreteras," *Excélsior*; Waters, "Remapping the Nation," 41–42.

56. Secretaría de Obras Públicas, *Caminos y Desarrollo*, 43; John Welsh, "Good Roads Lead into the Mexican Republic," *New York Times*, 24 February 1929; telegrams and letters on Calhoun Highway Association in, FP Obregón-Calles, 1921–28, vol. 27, files 104-C-60:104-C-139, AGN.

57. Alfonso Nava Negrete, "Derecho de las Obras Públicas en México," Biblio- teca Jurídica Virtual del Instituto de Investigaciones Jurídicas, Universidad Nacional Autónoma de México, http://biblio.juridicas.unam.mx/libros/6/2688/16.pdf (accessed 22 April 2015).

58. Ley de Caminos y Puentes, Adalberto Tejeda, ley 326, 26 July 1930, http:// sistemas.cgever.gob.mx/2003/Normatividad_Linea/constitucion_codigos_y_leyes/ley %20de%20caminos%20y%20puentes.pdf (accessed 18 March 2013).

59. Huasteca Petroleum, box 2145, exp. 57959, Veracruz, AHP.

60. Miguel W. Guerrero, chief litigator for HPC, complaint against government of Veracruz, 10 October 1930, FL, 1930–35, box 86, AGEV.

61. State response to HPC, 18 October 1930; order to Attorney General of Veracruz, 17 November 1930; Third District Court, 5 January 1931, FL, 1930–35, box 86, AGEV.

62. AHSCJN, files: Agwi 1 Incidente-743–1928; *amparo*, 16 March 1921, Huasteca; *amparo*, 17 November 1926, Huasteca; *amparo*, Cía. Internacional de Petróleo y Oleoductos, 4 March 1938; *amparo*, Ortiz Arnulfo, 6 March 1937; *amparo*, Román Martín, 7 March 1931.

63. "Giant Corporation Has Headquarters in City," 17 March 1932 and "Bank Chiefs Are Elected," *Brownsville Herald*, 14 January 1932.

64. "Mexico Begins Road Building," *Los Angeles Times*, 10 May 1925; "Como harán la construcción de las carreteras," *Excélsior*; Plutarco Elías Calles, state of the union address, 1 September 1925, IP, CD, DSI (Mexico City), 73.

65. Ocampo Bolaños, letter to President Ortiz Rubio, 2 February 1932, No. 501, FP Abelardo L. Rodríguez, vol. 217, exp. 573.1/8–573.3/26, 1932–34, Mexico City, AGN.

66. Bolaños, letter to Ortiz Rubio, 19 April 1932, No. 1771; Bolaños, confidential memorandum to Ortiz Rubio, 24 May 1932, No. 2367, FP Abelardo L. Rodríguez, vol. 217, AGN.

67. Bolaños, letter to Ortiz Rubio, 29 April 1932, No. 1796, FP Abelardo L. Rodríguez, AGN; "Creager Flies to Mexico on Business," *Brownsville Herald*, 28 April 1932.

68. Leopoldo Farías, Sub-Director of the Dirección Nacional de Caminos, letter to Guerrero, 22 June 1932, No. 1062.2735; Louis Swed, letter to Guerrero, 13 June 1932; SCOP, letter to Director General of the Dirección Nacional de Caminos (DNC), 22 June 1932, No. 13/2735, FP Abelardo L. Rodríguez, vol. 217, AGN.

69. Swed, letter to Miguel Acosta, 6 June 1932; Swed, letter to SCOP, 8 June 1932, FP Abelardo L. Rodríguez, vol. 217, AGN.

70. Creager and Bolaños, telegram to Ortiz Rubio, June 1932; Creager and Bolaños, letter to Ortiz Rubio, 17 June 1932, No. 13/2735, FP Abelardo L. Rodríguez, vol. 217, AGN.

71. Memorandum to SCOP (Swed accuses the agency of not acting in good faith), 3 September 1932, No. 368; Acosta, letter to Ortiz Rubio, 3 September 1932 (defends the agency's decisions), No. 362; Guerrero, telegram to SCOP (states that all communication with Swed has been turned over to the agency), 3 September 1932, FP Abelardo L. Rodríguez, vol. 217, AGN.

72. Niblo, *Mexico in the 1940s*, 253.

73. After Amasco's collapse, Creager and Swed returned to Brownsville to pursue new business interests; in 1934 Swed became general manager of a regional casino. See "Casino Moved," *Brownsville Herald*, 1 January 1934.

74. Abelardo L. Rodríguez, II Informe, 1 September 1934, IP, CD, DSI, Mexico City (2006), 113–14.

75. Nemesio García Naranjo, "México y el turismo," *El Porvenir*, 28 June 1936.

76. "Las buenas carreteras harán aumentar el automovilismo," *Excélsior*.

2. "Everyone Was Ready to Do Their Part"

The quote in the chapter title is from *Boletín de la Junta de Mejoras Materiales*, No. 1, 23 August 1926, SEGOB, Fomento, 1925–1930, Mejoras y Obras Públicas, Box 212, AGEV.

1. The term *flivver* is a colloquialism for the Ford Model T of 1908. Later it identified any car that was considered old and inexpensive. "Excursión de Veracruz a Jalapa en Autos Overland," *El Universal*, 16 July 1922; John Welsh, "Good Roads Lead into the Mexican Republic," *New York Times*, 24 February 1929.

2. Plutarco Elías Calles, fourth state of the union address, 1 September 1928, IP, CD, DSI, Mexico City (2006), 276.

3. Aboites and Loyo, "La construcción del nuevo Estado," 596–626.

4. Emilio Portes Gil, first state of the union address, 1 September 1929, IP, CD, DSI, Mexico City (2006), 56; Secretaría de Obras Públicas, *Caminos y Desarrollo*, 55.

5. DCOP, folder: "Informa a este Gobierno que los vecinos de varias municipales se rehúsan a prestar sus servicios voluntarios para hacer las mejoras en los caminos municipales," 24 September 1930, no. 55, fef. II.0.226, COP, box 101; DCOP, "Construcción carretera [*sic*] Altotonga-Tlapacoyan, conservación de la misma, gastos relativos, etc.," 2 January 1933, no. 26, ref. II.0.226, 1934, SEGOB, SCOP, box 64, AGEV; "Los transportes que actúan legalmente, al señalar anomalías de los que operan al margen de la misma como 'piratas,' solo defienden su derecho legítimo," 1 August 1952, and "Con la socorrida arma de los amparos siguen luchando los camioneros piratas," *El Norte*, 30 August 1952.

6. "Plan para financiar la construcción y conservación de un sistema de caminos nacionales," 28 October 1932, FPAbelardo L. Rodríguez, vol. 217, files 573.1/8 to 573.3/26, 1932–34, AGN.

7. Secretaría de Obras Públicas, *Caminos y Desarrollo*, 44–45; "Informe al C. Presidente de la Republica de los trabajos ejecutados por la Comisión Nacional de Caminos durante el mes de mayo 1928," 24 July 1928, FP Obregón-Calles, vol. 27, files 104-C-60 to 104-C139, AGN.

8. "Informe al C. Presidente," files 104-C-60 to 104-C139, AGN; "Como harán la construcción de las carreteras," *Excélsior*, 12 August 1925.

9. Ciénaga de Flores, "Piden garantías de que se construyan las fincas que se van a derrumbar para la construcción de la Carretera Nacional," 1929, Fomento; Monterrey, "Oficio en el cual se les notifica que deben cumplir con las disposiciones de la Comisión Nacional de Caminos," 1928, Fomento, Comisión Nacional de Caminos, 1921–59, box 1, AGGNL.

10. "Caminos de Nuevo León. Detalle del movimiento general," 31 August 1931, JLCNL (SCOP), 1931, box 5, AGGNL.

11. Comisión de Caminos del Estado (CCE), "Lista de rayas de las semanas no. 1 a la no. 26," 1931, JLCNL (SCOP), 1931, box 4, AGGNL.

12. R. Garcia Alba, manager of Compañía Constructora Anáhuac, S.A., letter to José Benitez, governor of Nuevo León, requesting delivery of motor vehicles, 12 January 1928; Benitez, letter to Alba, authorizing 75,000 pesos for bridge construction, 14 April 1928; Benitez, letter to Adolfo Larralde Jr., on bridge from Colonia Independencia to the rest of the city, 14 April 1928, Comisión Nacional de Caminos Vecinales, 1921–59, box 1, AGGNL; Waters, "Remapping Identities," 223.

13. Porfirio Treviño, whom Anáhuac listed as a consulting engineer in 1928, reappears as "jefe de ingenieros" (head of engineers) for the JLCNL in 1933. See Alba, letter to Benitez, 12 January 1928 and Francisco Cárdenas, "Programa de trabajos para el mes de septiembre 1933," 15 August 1933, JLCNL (SCOP) 1933, box 7.

14. "Lo relativo a la apertura de la calle que parte del Dique y termina en la Avenida Bolívar y todo lo demás que se relaciona con la misma," 14 May 1926, folder 7-3.552, Fomento, 1925–30, box 206, AGEV.

15. "Cooperación de esa ayuntamiento para la construcción del camino México-Jalapa," 3 April 1930, no. 1624, exp. 2.002–9, SEGOB, COP, box 1, AGEV.

16. Knight, "The Characters and Consequences of the Great Depression in Mexico," 219.

17. Secretaría de Obras Públicas, *Caminos y Desarrollo*, 44; Toscano, letter to the DNC confirming Governor Pablo Quiroga as the new head of the JLCNL, 28 December 1933, AJLCNL, box 7. In 1933 Salvador Toscano appears in the state archives as "Representante de la Secretaría de Comunicaciones y Obras Públicas," and correspondence from this period indicates he supervised the technical aspects of the JLCNL, for example: Toscano, letter to Guillermo Aguilar Álvarez, chief of the Departamento de Cooperación, 19 January 1934 (discussing progress on the Linares-Matehuala and El Jabalí-Monterrey–Los Muertos roads), JLCNL, box 10, AGGNL.

18. Secretaría de Obras Públicas, *Caminos y Desarrollo*, 47; files 2.4.00.43, 1.10.00.48, 9, 10.4, JLCNL (SCOP) 1934, box 10, AGGNL.

19. Miguel Alemán, Poder Ejecutivo, *Gaceta Oficial de Veracruz*, 20 February 1937, exp. 317/6, 1939, box 619, AGEV.

20. Toscano, letter to the DNC confirming Governor Pablo, 28 December 1933, JLCNL, AGGNL.

21. For an example of a typical request see Toscano, letter to Álvarez for disbursal of road funds, 27 July 1934, JLCNL, 1934, box 10; Toscano, letter to Álvarez for payment of engineers' salaries, 25 May 1934, JLCNL (SCOP) box 9, AGGNL.

22. Toscano, letter to Álvarez for disbursal of road funds, 27 July 1934, JLCNL, 1934, box 10; Toscano, letter to Álvarez for payment of engineers' salaries, 25 May 1934, JLCNL (SCOP) box 9, AGGNL.

23. Carreteras de México, S.A. July 27 de 1943, vol. 136, no. 9006, F. 264, Acervo Histórico del Archivo General de Notarías, Mexico City; Efraim Zadoff, "Sourasky," *Encyclopaedia Judaica*, http://www.encyclopedia.com/religion/encyclopedias-almanacs-transcripts-and-maps/sourasky (accessed 14 January 2017). For more on Platt, see Gómez Estrada, *Gobierno y casinos*, 130–31, 136–37. Gómez Estrada does not label Platt a formal *prestanombre* (nominal holder) of Rodríguez, but rather identifies him as an individual who used to funnel money to political allies and invest in economic interests on Rodríguez's behalf.

24. CCE de Nuevo León, "Red de Caminos" (Road Network), map, 1932, JLCNL (SCOP) 1932, box 6, AGGNL; U.S. Bureau of Public Roads, "Inter-American Highway: Map of Tentative Route," October 1933, Washington, DC, https://commons.wikimedia.org/wiki/File:Inter-American_Highway_map_October_1933.jpg (accessed 14 January 2017).

25. "Informando sobre carreteras construidas y en construcción de este estado," 29 January 1931, exp. S25.2.003-1, 1931, SEGOB, COP, box 27, AGGNL.

26. CCE de Nuevo León, "Red de Caminos," 1932.

27. Toscano, letter to the governor regarding six-year plan for roads, 11 August 1933; Toscano, letter to the DNC on road program in Nuevo León, 2 August 1933; Toscano, letter to various public officials on guidelines for road building, 27 July 1933, JLCNL (SCOP) 1933, box 7, AGGNL.

28. Francisco Cárdenas, "Programa de trabajos para el mes de septiembre 1933," 15 August 1933, JLCNL (SCOP) 1933, box 7; CCE, "Informes Camino a Villa de García,"

JLCNL (SCOP) 1932, box 6; Pablo Quiroga, telegram to Francisco Cárdenas, 12 November 1932, Quiroga, telegram to Francisco Cárdenas, 12 November 1932, Comisión Nacional de Caminos, 1921–59, box 1, AGGNL.

29. CCE, "Informes Camino a Villa de García"; Toscano, letter to Charles Mumm, describing surface materials of Linares-Matehuala road, 3 March 1934, JLCNL (SCOP) 1934, AGGNL.

30. File: "Caminos y Carreteras," book I, D/661 (5-3)/10, Comisión Nacional de Caminos, 1921–59, box 1; Cárdenas, "Programa de trabajos para el mes de septiembre 1933," box 7; Toscano, "Programa de trabajos para el mes de marzo 1934," 1 March 1934, JLCNL (SCOP) 1934, box 9; JLCNL, "Camino Matamoros-Mazatlán," JLCNL (SCOP) 1933 box 7; Toscano, "Presupuesto de gastos para el mes de junio 1934," 1 June 1934 and "Presupuesto de gastos para el mes de marzo de 1934," 1 March 1934, JLCNL (SCOP) 1934, box 9, AGGNL.

31. CCE, "Informes Camino a Villa de García," JLCNL (SCOP) 1932, box 6; Toscano, letter to various public officials on guidelines for road building, 27 July 1933, JLCNL (SCOP) 1933, box 7, AGGNL.

32. Cárdenas, "Programa de trabajos para el mes de septiembre 1933"; unsigned letter to DNC on road damage due to hurricane weather near Linares, 7 July 1933, JLCNL (SCOP) 1933, box 7, AGGNL.

33. Comité Pro-Abastecimiento de Agua, letter to the Secretario de Comunicaciones y Obras Públicas, 13 August 1942, JLCNL (SCOP), box 33, AGGNL. The letter highlights the positive economic impact the road had on the area in the 1930s, while urging the government to repair the road due to the lack of proper maintenance over the past decade. Toscano, letter to commission supervisors, 26 March 1934, JLCNL (SCOP) box 9; U.S. Bureau of Public Roads, "Inter-American Highway: Map of Tentative Route," October 1933; Joaquin Garza y Garza, letter to the Governor of Nuevo León, 24 June 1933, and Toscano, letter to the governor, 14 July 1933, JLdCNL (SCOP) 1933 box 7, AGGNL.

34. "Informando sobre carreteras," 29 January 1931, box 27; Waters, "Remapping the Nation," 87–88; Fowler Salamini, *Agrarian Radicalism in Veracruz*, xiv–xv, 25–29.

35. "Informando sobre carreteras," 29 January 1931, box 27. For tax discussion and negotiations in Banderilla, see folder 29, correspondence from April 1930 to 12 August 1930, SEGOB, COP, box 1, AGEV.

36. "Informando sobre carreteras."

37. María Luisa Esteva, letter to Judge of First Instance of the Judiciary in Xalapa, 29 October 1930, exp. 2.069, COP, 1930–35, box 86, AGEV.

38. Luis Monterogüido, Judge of the First Instance, letter to Governor Tejeda, 4 November 1930; Governor Tejeda, letter to Monterogüido, 5 November 1930; Arturo Martínez, First District Judge of Veracruz, ruling on the Esteva case, letter to Governor Tejeda, 15 November 1930, exp. 2.069, COP, 1930–35, box 86, AGEV.

39. "Informando," 29 January 1931, box 27; Secretaria de la Economía Nacional, Dirección General de Estadística, *Quinto Censo de Población: Estado de Veracruz*, 15 May 1930 (Mexico City), http://www3.inegi.org.mx/sistemas/biblioteca/ficha.aspx?upc=702825411718 (accessed 14 January 2017).

40. DCOP, folder: "Solicita que el camino de Huatusco a Puente Nacional pase por

Tlacotepec, Ver.," no. 22, ref. II.o.226, 15 June 1934, SEGOB, COP, 1934, box 66, AGEV; Waters, "Remapping the Nation," 98–99; Haen, *Fields of Power, Forests of Discontent*, 84.

41. DCOP, folder: "Solicita que el camino de Huatusco a Puente Nacional . . ."; Waters, "Remapping the Nation," 98–99.

42. DCOP, folder: "Informa a este Gobierno que los vecinos de varias municipales se rehúsan a prestar sus servicios voluntarios para hacer las mejoras en los caminos municipales," 24 September 1930, no. 55, ref. II.o.226, COP, box 101, AGEV; Waters, "Remapping the Nation," 99.

43. DCOP, folder: "Lo relacionado con la ayuda que solicitan para que pueda llevar a cabo el Presidente Municipal de aquel lugar, los trabajos del camino carretero," 24 June 1931, no. 59, ref. II.o.226, SEGOB, COP, box 25, AGEV.

44. DCOP, folder: "Todo lo relacionado con la petrolización del Camino Carretero Cordoba-orizaba," May 1934, no. 7, ref. II.2.002, SEGOB, COP, box 63; DCOP, folder: "Solicita que el camino de Huatusco a Puente Nacional," box 66; Alfonso Romero, letter to the governor, 28 February 1941 (request for tools to make road repairs), exp. 317.42, box 834, AGEV; Secretaria de la Economía Nacional, Dirección General de Estadística, *Quinto Censo de Población: Estado de Veracruz*, 15 May 1930 (Mexico City).

45. DCOP, folder: "Todo lo relacionado con la petrolización del Camino Carretero Cordoba-orizaba," May 1934; DCOP, folder: "Solicita que el camino de Huatusco a Puente Nacional," box 66; Alfonso Romero, letter to the governor, 28 February 1941; *Quinto Censo de Población: Estado de Veracruz*.

46. Surviving documents in the state archives are incomplete and do not provide clear figures on monthly progress, making it difficult to calculate unit costs per cubic meter of highway built. DCOP, "Construcción carretera [sic] Altotonga-Tlapacoyan, conservación de la misma, gastos relativos, etc.," 2 January 1933, no. 26, ref. II.o.226-, 1934, SEGOB, SCOP, box 64, AGEV.

47. DCOP, "Construcción carretera [sic] Altotonga-Tlapacoyan, conservación de la misma, gastos relativos, etc."

48. DCOP, "Construcción carretera [sic] Altotonga-Tlapacoyan, conservación de la misma, gastos relativos, etc."

49. Wendy Waters and J. Brian Freeman, as well as Stephen J. Batchelor and Myrna Santiago, among others, have discussed the importance of modern labor arrangements to national and state politicians in Mexico as a key component of state formation in the late nineteenth and twentieth centuries.

50. DCOP, folder: "Construcción, mejoramiento, gastos relativos de la carretera Cardel a San Carlos," 2 January 1933, no. 13, ref. II.2002, COP, box 72, AGEV.

51. Harry Skipsey, mill operator, letter to Governor Salcedo Casas, 17 July 1933, no. 13, ref. II.2002, 1934, COP, box 72, AGEV.

52. Eulalio Gutiérrez, letter to Governor Gonzalo Vásquez Vela, 18 July 1933, no. 25, ref. II.0026, SEGOB, COP, box 67, AGEV.

53. Estimated exchange rate for Mexican pesos to U.S. dollars in the 1930s was, on average, 4.5 to 1; in 2014 dollar amounts, these penalties range between $15 and $2,700 (author's estimates). See Nacional Financiera, *50 años de revolución en cifras*; "Lo relacionando con las multas impuestas por infracción la Ley de Tránsito en el Estado" [sic], file, 2 August 1930, box 129 (1921–35), COP, AGEV; SEGOB, DCOP, ref. II.O.34,

no. 6, Presidente Municipal de Jalapa, Asuntos relativos a quejas del Ayuntamiento por infracción al Reglamento de Tráfico, rile 273, AGEV.

54. File: Comunicaciones, Transito, Solicitudes, Córdoba, Ver., Unión de Choferes, March 1933, no. 5, ref. II 0.32, box 44, FCOP, AGEV.

55. File: Comunicaciones, Transito, Solicitudes, Córdoba, Ver., Unión de Choferes.

56. The government may have underestimated the number of vehicles carrying passengers; although federal officials counted automobiles, buses, and trucks separately, documents indicate that some transport cooperatives purchased trucks and converted them for regional passenger service. Ford Motor Company, "Rebaja de Precios" (advertisement), *El Dictamen*, 8 November 1933; "Sección de Carreteras," July–August 1936, book 2, no. 12, *Correos y Transportes*, box 346, carreteras, AGEV; Nacional Financiera, *50 años de revolución en cifras*, graph "Vehículos de motor en circulación," 100; Instituto Nacional de Estadística y Geografía (National Institute of Stastistics and Geography, INEGI), graph "Volumen y crecimiento," in the statistical compendium "Población total por entidad federativa," 1930, online.

57. Freeman, "'Los hijos de Ford,'" 218–21.

58. Antonio Alvarez et al., letter to Governor Alemán, 9 December 1936; Casas Alemán, letter to Antonio Alvarez et al., 6 January 1937, exp. 196/1 114, box 340, 1937, AGEV.

59. "Bases Constitutivas," Sociedad Cooperativa de Camioneros del Servicio Urbano de Jalapa y Anexas, 1934, box 346, 1937, AGEV.

60. "Bases Constitutivas."

61. File: "Comunicaciones, Tránsito, Permisos," Jalapa, Ver., Angel Rivera, June 1933, no. 11, ref. II.0.32; file: "Comunicaciones, Tránsito, Permisos," Córdoba, Ver., Eustacio Paul and Humberto Calvo, May 1933, no. 7, ref. II 0.32, box 44, FCOP, AGEV.

62. "Modificaciones al sistema de transito en la ciudad," *El Porvenir*, 23 June 1936. For more information on the history of Monterrey's transit laws, see my article, "'Neither Motorists nor Pedestrians Obey the Rules.'"

63. Freeman, "'Los hijos de Ford,'" 219–20.

64. "Bases Constitutivas," Sociedad Cooperativa de Camioneros del Servicio Urbano de Jalapa; file: "Comunicaciones, Tránsito, Permisos," Jalapa, Ver., AGEV.

65. File: Cooperativa de Auto-Transportes, S.C.I. México, February 1933, no. 1, ref. II 0.34, box 103, 1930–35, FCOP, AGEV; "Los transportes que actúan legalmente . . . ," 1 August 1952, and "Con la socorrida arma . . . ," *El Norte*, 30 August 1952.

3. "So That These Problems May Be Placed"

The quote in the chapter title is from Silvano Reyes, ejidal committee president, document describing formation of Ejidatarios del Rio de Tecolapan, 17 August 1937, exp. 317/1, box 343, AGEV.

1. "Ceremonia de la inauguración de la carretera," *El Porvenir*, 28 June 1936; U.S. Department of State, Records of the Department of State Relating to the Internal Affairs of Mexico, 1930–39, *Transmission of Publication Issued by the Mexican Department of Labor*, 1985, microfilm, roll 5, National Archives, Washington DC; "Daniels Dedicates Plaque in Mexico," *New York Times*, 5 July 1936.

2. "El movimiento en las líneas de camiones," *El Universal*, 29 March 1934.

3. Aboites and Loyo, "La construcción del nuevo Estado," 623–46.

4. Fulwider has also noted the nuances of Cárdenas's road-building policies that shared commonalities with those of his predecessors and were also useful in reaching out to some conservatives and businessmen ("Driving the Nation," 61–86).

5. Abelardo L. Rodríguez, II Informe, 1 September 1934, IP, CD, DSI, Mexico City (2006), 113–14.

6. "Mexico to Speed Road Building," New York Times, 11 March 1937.

7. "Mexico Plans 75% Reduction in Debt," New York Times, 23 December 1936.

8. Harold Callenders, "A Big Rush to Mexico," New York Times, 21 July 1935

9. "El transito sobre la carretera Monterrey-Reynosa ha aumentado," El Porvenir, 1 August 1937.

10. Francis Brown, "Mexico's Season On," New York Times, 3 January 1937; Russell Franck, "Good Roads in Mexico," New York Times, 3 September 1939; Nacional Financiera, 50 años de revolución mexicana en cifras, 100.

11. Lázaro Cárdenas, IP, CD, DSI, Mexico City (2006); "Las carreteras en México," Eco Xalapeño, 10 December 1937.

12. Compañía Mexicana de Garantías, Loan document, 15 June 1936, file: 1.2.5 Fianzas y Garantías, box 19, 1938, JLCNL, AGGNL, Monterrey.

13. Pablo Domínguez, letter to Arnulfo Canales, 8 October 1939, file: 1.2.5 Fianzas y Garantías, box 19, 1938, JLCNL, AGGNL.

14. Miguel Alemán, Poder Ejecutivo, Gaceta Oficial de Veracruz, 20 February 1937, exp. 317/6, 1939, box 619, AGEV.

15. Compañía Internacional Petróleo y Oleoductos, S.A., amparo, 4 March 1938, Semanario Judicial, Suprema Corte de la Justicia de la Nación (SCJN).

16. Terreros Fermín, amparo, 31 March 1936; Bonilla Benjamín, amparo, 1 October 1936, Seminario Judicial, SCJN.

17. Ruíz Guadalupe, amparo, 30 September 1939.

18. Secretaría de Comunicaciones y Transporte, Historia de las Juntas Locales de Caminos, 408.

19. Domínguez, "Programa de trabajos para el año 1937 en cooperación con el Gobierno Federal," 1 December 1936, box 14, 1936, JLCNL, AGGNL; JLC de Orizaba, "Detalle de trabajos desarrollados"; Soto, budget report to Miguel Alemán, no. 935, exp. 317/41, AGEV.

20. State officials chose the name Junta Central de Caminos de Veracruz to avoid confusion with the Junta Local de Caminos de Orizaba, in operation since 1929. See Secretaría de Comunicaciones y Transporte, Historia de las Juntas Locales de Caminos, 290.

21. Secretaría de Comunicaciones y Transporte, Historia de las Juntas Locales de Caminos, 215.

22. Manuel Saldaña, Sindicato de Empleados y Obreros Constructores de Caminos en el Estado de Nuevo León, 1942, serie: 103, file: 16 Sindicato, box 34; Salvador Toscano, letter to director of Hospital González, 28 June 1934, exp. Minutario Enero a Junio, box 9, 1934, JLCNL, AGGNL.

23. Domínguez, letter to Guillermo Aguilar Álvarez, 5 September 1936, exp. Aportaciones del año, box 14, 1936, JLCNL, AGGNL; Secretaría de Comunicaciones y Transporte, Historia de las Juntas Locales de Caminos, 212–16.

24. Domínguez, letter to Rafael Rodríguez, secretary general of SEOCC, 17 February 1938, exp. 1.7-3 16-2, exp. Asuntos con el Sindicato de Empleados y Obreros Constructores de Caminos en el Estado de Nuevo León, box 19, 1938, JLCNL, AGGNL.

25. Secretaría de Comunicaciones y Transporte, *Historia de las Juntas Locales de Caminos*, 212–16; Leopoldo Banda, letter to Carlos Barney, 30 March 1938, exp. LB-22 16-2, box 19 SEOCC; letter to Domínguez, 19 May 1938, box 19, 1938, AGGNL.

26. "Personal Administrativo y de Campo," January 1939, exp. 1.7.3.2.1, box 23, 1939.

27. Domínguez, letter to the Asociación de Beis-bol de Monterrey, 14 March 1938, JLCNL (SCOP), box 2, 1938, AGGNL; Secretaría de Comunicaciones y Transporte, *Historia de las Juntas Locales de Caminos*, 212–16.

Steven Bachelor finds similar strategies at work in the 1950s and 1960s as U.S. automobile manufacturers sponsored sports teams and organized other kinds of community events for workers and their families to forge ties with employees that extended beyond the shop floor as a means to reinforce company loyalty and defuse tensions with management. See "Toiling for the 'New Invaders,'" 285–308.

28. Workers' complaint to Domínguez, undated, box 19, 1938, JLCNL, AGGNL.

29. Gobierno de Nuevo León, *Caminos, 1956–1959*, map, "Caminos Construidos" (Monterrey: Departamento de Prensa, 1959); "Mañana se inaugura la carretera," *El Porvenir*, 18 July 1941; Saragoza, *The Monterrey Elite and the Mexican State*, 240–50. Compañía Fundidora de Fierro y Acero de Monterrey (Monterrey Iron and Steel Foundry) was one of the most important firms in the region during the early and mid-twentieth century. Although much of the company has been sold and its facilities dismantled as of 2013, one of its main foundries, known as Horno 3, stands as an iconic symbol of the city, serving as a museum and centerpiece in Monterrey's popular Foundry Park.

30. A. W. Villarreal, letter to the governor, 6 August 1937, exp. D/021 (06)/3742, box 1, Comisión Local de Caminos (CLC), 1867–52, AGGNL.

31. Ramiro Tamez, letter to Villareal stalling construction proposal, exp. D/21(06)/3742, 13 August 1937, box, CLC, AGGNL.

32. Salvador Toscano, SCOP contract with Carreteras de México, 1935, exp. 18 Compañía Construcciones Nacionales, exp. 18.1 Contrato, box 26, 1940, JLCNL, AGGNL.

33. Exp. 18.4, Construcciones Nacionales, Estimaciones Mensuales, July–December 1940, box 26, 1940; exp. 18.2 Personal; 18.3 Herramienta y Equipo Alquilado por la Junta Local, box 26, 1940, JLCNL, AGGNL.

34. Gilberto del Arenal, letter to Domínguez, 26 September 1940, exp. 18.1, box 26, 1940, AGGNL.

35. Prospero Castro, letter to Domínguez, 24 October 1940; Domínguez, letters to Castro and Tomás Williams, 28 October 1940, exp. 18.1, box 26, 1940, AGGNL.

36. Domínguez, letter to Williams, 20 September 1940, exp. 18.1, box 26, 1940, AGGNL.

37. Bonifacio Salinas de Leal, governor of Nuevo León, letter to Armando Salinas, head of Department of Cooperation, DNC, SCOP, 6 August 1941, exp. 18.1, box 26, 1940, AGGNL.

38. Mario Ojeda, letter to Rafael Colombón, head of DCOP, 17 July 1936, exp. 317/6, 1939, box 617, AGEV; Secretaría de Comunicaciones y Transporte, *Historia de las Juntas Locales de Caminos*, 293.

39. Mario Ojeda, letter to Rafael Colombón, head of DCOP, 17 July 1936, exp. 317/6, 1939, box 617, AGEV; Secretaría de Comunicaciones y Transporte, *Historia de las Juntas Locales de Caminos*, 293.

40. Miguel Cataño Morlet, letter to Luis Niño Castillo, president of the Comité de Caminos de Coatzacoalcos, 29 November 1938, exp. 317/49, box 619, 1939, AGEV; Gaceta Oficial de Veracruz, 20 February 1937, book 37, no. 22; "Un cabo de trabajadores de la Carretera Jalapa–Las Vigas fue acusado de rapto," 30 November 1937; "Cuatro sujetos de los que trabajan en la carretera violaron a una niña," 10 December 1937; "Fue reorganizado el personal de la Junta Central de Caminos," 14 December 1937; "El Ing. Cataño Morlet renunció como Representante del Gobierno ante la Junta Central de Caminos," 18 December 1937, *Eco Xalapeño*.

This departure from public service, however, was short-lived; months later, in 1938, he reappeared in official correspondence as Veracruz's director general for road building, involved with planning efforts for a new motorway from Minatitlán to Coatzacoalcos. See Cataño Morlet, letter to Luis Niño Castillo, president of the Comité de Caminos de Coatzacoalcos, 29 November 1938, exp. 317/49, box 619, 1939, AGEV; Gaceta Oficial de Veracruz, 20 February 1937, book 37, no. 22.

41. JLC de Orizaba, "Detalle de trabajos desarrollados por esta Junta Local de Caminos comprendidos desde el 17 Septiembre 1934 hasta la fecha," AGEV.

42. JLC de Orizaba, "Detalle de trabajos desarrollados por esta Junta Local de Caminos comprendidos desde el 17 Septiembre 1934 hasta la fecha," AGEV.

43. Cataño Morlet, letter to Castillo, 29 November 1938, exp. 317/49; "Lista de raya," Comité Pro-Camino Puerto Mexico-Minatitlan, 7–13 October 1938, box 619, 1939, AGEV.

44. Petitions from ejidal committees in Zongolica, exp. 115/112, 1938; Adolfo Omaña, letter to Alejando Montalvo, 28 November 1938, box 4, 1937–41, Asuntos Indígenas, AGEV.

45. Agustín Flores, secretary general of the Confederación Sindicalista de Obreros y Campesinos del Estado de Veracruz (Labor Confederation of Workers and Farmers of Veracruz, CSOCV), letter to Governor Vázquez Vela, 16 July 1934, exp. 317/10.114, 1936, box 200, AGEV; Martín Torres, members of CROM, letter to Governor Vázquez Vela, 1 April 1934, exp. 317/10.114, 1936, box 200, AGEV; Waters, "Remapping the Nation," 155–16.

46. "Los campesinos . . . ,"*La Voz del Campesino*, 25 August 1937; Luis Sosa, president of the Unión de Auto-Transportes de Córdoba, letter to Governor Vázquez Vela, 17 July 1934, exp. 317/10.114, 1936, box 200, AGEV.

47. "Los campesinos . . . ,"*La Voz del Campesino*, 25 August 1937; Luis Sosa, president of the Unión de Auto-Transportes de Córdoba, letter to Governor Vázquez Vela, 17 July 1934, exp. 317/10.114, 1936, box 200, AGEV.

48. Flores, CSOCV, letter to Vázquez Vela, 16 July 1934, exp. 317/10.114, 1936, box 200; Federación de Trabajadores del Estado de Veracruz, letter to the governor, 16 April 1941, exp. 317/54, 1943, box 1059, AGEV.

49. Nacional Financiera, *50 años de revolución mexicana en cifras*, 100; INEGI, *México en cifras, Veracruz*, http://www.beta.inegi.org.mx/app/areasgeograficas/?ag=30 (accessed 14 January 2017); "El movimiento en las líneas," *El Universal*.

50. SCOP, "Administración de la carretera Xalapa-Veracruz," exp. 515.1/42.9, vol.

630, Serie Ferrocarriles, exps. 515.1-169-515.1-251, 1934–40, FP Lázaro Cárdenas, AGN; Autobuses de Oriente, "Viaje Cómodo y Seguro," advertisement, *El Tema de Hoy*, 3 January 1949; Transportes Frontera, "Viajes de México a Monterrey y Laredo," advertisement, *Excélsior*, 20 April 1939.

51. SCOP, "Administración de la carretera Xalapa-Veracruz," exp. 515.1/42.9,; Autobuses de Oriente, "Viaje Cómodo y Seguro," advertisement, *El Tema de Hoy*; Transportes Frontera, "Viajes de México a Monterrey y Laredo," advertisement, *Excélsior*.

52. Enrique Rosero, Transportes Huatusco, letter to President Cárdenas, 13 May 1936, exp. 515.1/180, exp. 515.1/42.9, vol. 630, Serie Ferrocarriles, exps. 515.1-169-515.1-251, 1934–40, FP Lázaro Cárdenas, AGN; Nacional Financiera, *50 años de revolución mexicana en cifras*, 100.

53. Enrique Rosero, Transportes Huatusco, letter to President Cárdenas, 13 May 1936; Nacional Financiera, *50 años de revolución mexicana en cifras*, 100.

54. "Mexican Oil Strike Called," *New York Times*, 19 May 1937; "Mexico Oil Strike Strands Tourists," *New York Times*, 29 May 1937; "Mexico Is Gripped by Walkout Wave," *New York Times*, 30 May 1937.

55. "Mexican Oil Strike Called," *New York Times*; "Mexico Oil Strike Strands Tourists," *New York Times*; "Mexico Is Gripped by Walkout Wave," *New York Times*.

56. Discurso del Presidente Lázaro Cárdenas con motivo de la Expropiación Petrolera, delivered 18 March 1938. http://www.biblioteca.tv/artman2/publish/1938_227/Discurso_del_Presidente_L_zaro_C_rdenas_con_motivo_1442.shtml (accessed 6 July 2016); Frank Kluckhohn, "Fascist Influence Growing in Mexico; U.S. Trade Suffers," *New York Times*, 15 August 1938.

57. Nacional Financiera, *50 años de revolución mexicana en cifras*, 146.

58. "Memorial que presentaron las cooperativas y empresas de camiones al honorable ayuntamiento," *El Porvenir*, 1 March 1942.

59. Cárdenas, IP, CD, DSI, 224–30; Manuel Ávila Camacho, first state of the union address, 1 September 1941, IP, CD, DSI, 62.

4. "We March with Mexico for Liberty!"

The quote in the chapter title is from *Gaceta Oficial del Estado de Veracruz*, 8 December 1942, 12.

1. Jorge Cerdán, letter to Manuel Ávila Camacho, 1942, box 958, AGEV.

2. Secretaría de Obras Públicas, *Caminos y Desarrollo*, 59; Manuel Ávila Camacho, first state of the union address, 1 September 1941, IP, CD, DSI, Mexico, DF (2006), 62.

3. Comité Pro-Abastecimiento de Agua, letter to the Secretario de Comunicaciones y Obras Públicas, 13 August 1942, box 33, JLCNL, SCOP, AGGNL; José Peña Avila, letter to the JLCNL, 20 June 1945, JLCNL, box 40; letter from residents of Paso de la Loma to Domínguez, 9 July 1945, JLCNL, box 40, AGGNL.

4. Pedro Martínez Tornel, "Memoria de la Secretaría de Comunicaciones y Obra Públicas," September 1945–August 1946, SCOP, Mexico City; Ávila Camacho, first state of the union address, IP, CD, DSI, Mexico, DF (2006), 45.

5. Frank Kluckhohn, "U.S. and Mexico in Wide Pact Set Oil Procedure," *New York Times*, 19 February 1941; "Comentase en Estados Unidos el pacto que se firmó ayer," *Excélsior*, 20 November 1941.

6. Niblo, *War, Diplomacy, and Development*, 74–76; "Mexico Halts Deals for Delivery of Oil," *New York Times*, 30 August 1939; "Mexico Cancels a Concession," *New York Times*, 23 October 1940.

7. "Monterrey espera hoy la caravana fúnebre del Portero del Llano," *El Norte*, 21 May 1942; "Monterrey vistió de luto," *El Norte*, 22 May 1942; *El Tiempo de México*, May 1942–May 1943; "Suplemento del año nuevo," *El Tiempo*, 8 January 1943.

8. Gobierno del Estado de Veracruz, "Un férvido homenaje al señor General de División Manual Ávila Camacho," *Excélsior*, 16 December 1942.

9. *Gaceta Oficial del Estado de Veracruz*, 8 December 1942, 12.

10. PEMEX-Penn, "Cuidar el auto o irse de pie," advertisement, *El Tiempo*, 29 January 1943.

11. Veedol, "He aquí el auto de 1943 que usted no compró," advertisement, *El Porvenir*, 2 April 1943.

12. "Colapso del turismo en México," *El Porvenir*, 7 March 1942.

13. Paz, *Strategy, Security, and Spies*, 218–19.

14. Domínguez, letter to Cisneros Oil Company, 30 March 1944, JLCNL, 1944, box 38, AGGNL.

15. Arturo B. de la Garza, letter to Maximino Ávila Camacho, 4 November 1944; Pedro Martínez Tornel, letter to Arturo B. de la Garza, 18 August 1944, JLCNL, box 38, AGGNL.

16. De la Garza, letter to Maximino Ávila Camacho, 11 May 1944, box 38, AGGNL.

17. "Auto Crisis" and "Refacciones para camión Chevrolet," *El Tiempo*, 18 December 1942; "Autotransportes" and "El Mercado negro," *El Tiempo*, 28 May 1943.

18. "Autotransportes" and "El Mercado negro," *El Tiempo*.

19. "Poder Judicial: Camiones Regiomontanos," *El Tiempo*, 22 January 1943.

20. "Se conjuro el paró de camioneros," *Excélsior*, 5 August 1944; "Autotransportes" and "El Mercado negro," *El Tiempo*.

21. Rubén Bauchez, letter to Cerdán, 28 July 1943, box 1060, exp. 317/70 Huiloapan-Nogales, AGEV.

22. Rubén Bauchez, letter to Cerdán, 28 July 1943.

23. The Municipal Anti-Nazi Fascist Committee, Martín Aldama, Comité Nacional Antinazifascista, letter to Governor Cerdán, 21 April 1943, exp. 232-2073, box 1059, AGEV.

24. Chamber of Commerce of Papantla, letter to Governor Cerdán, 14 March 1944, exp. 45-4607, box 1059, AGEV.

25. Krauze, *La presidencia imperial*, 43–45; Quintana, *Maximino Ávila Camacho and the One-Party State*, 110. It is noteworthy that Betancourt later became governor of Puebla.

26. Governor Bonifacio Salinas, telegram to President Ávila Camacho, 17 October 1942; Salinas, letter to the president, 22 March 1941; Waldo Romo Castro, personal aide to President Ávila Camacho, letter to SCOP (outlining road budget for Nuevo León), 18 December 1942, box 571, exp. 515.1/85-515.1/154, FP Ávila Camacho, AGN; Domínguez, letter to Sindicato de Empleados y Obreros Constructores de Caminos de Nuevo León (Union of Employees and Road Workers of Nuevo León, SEOCCENL), 31 March 1942; Manuel Lopez (SEOCCENL), letter to Domínguez, 29 March 1942, JLCNL, box 34, JLCNL, AGGNL.

27. "D. Maximino A. Camacho asumió su ministerio," *El Norte*, 3 October 1941.

28. "El nuevo ministro de comunicaciones," *El Norte*, 7 October 1941.

NOTES TO PAGES 101–107

29. "El nuevo ministro de comunicaciones," *El Norte.*

30. See "Seminario Judicial": *amparo*, 1 August 1941; *amparo*, 23 July 1941; *amparo*, 10 June 1942; *amparo*, 21 August 1942; *amparo*, 10 February 1944, AHSCJN.

31. See "Seminario Judicial": *amparo*, 1 August 1941; *amparo*, 23 July 1941; *amparo*, 10 June 1942; *amparo*, 21 August 1942; *amparo*, 10 February 1944, AHSCJN.

32. Domínguez Robles Gabriel, *amparo*, 10 June 1942, Segunda Sala; Garaby Cuevas María, *amparo*, 10 February 1944, Segunda Sala, "Seminario Judicial," AHSCJN.

33. Sindicato de Propiertarios de Auto-Camiones de la Linea México-Cuaula-Matamoros, Oaxaca y Anexas y coags. *amparo*, 21 August 1942, Segunda Sala; Gil Armando, *amparo*, 23 July 1941, Segunda Sala, "Seminario Judicial," AHSCJN.

34. Ortega Brito, Gregorio y coag., *amparo*, 1 August 1941, Segunda Sala, Seminario Judicial, AHSCJN.

35. Domínguez, letter recommendation for José Herrera to Cristales Mexicanos, S.A., box 40, JLCNL, AGGNL.

36. Manuel Saldaña, letter to Domínguez and Salinas, 24 November 1942, box 34, JLCNL, AGGNL.

37. Francisco Herrera, letter to Domínguez, 13 February 1943, box 38, JLCNL, AGGNL.

38. Próspero Castro, letter to Domínguez, 11 October 1940; Domínguez, letter to SCOP, 24 October 1940, box 26, JLCNL, AGGNL; Salinas, telegram to President Ávila Camacho, 17 October 1942; Salinas, letter to the president, 22 March 1941; Waldo Romo Castro, personal aide to President Ávila Camacho, letter to SCOP (outlining road budget for Nuevo León), 18 December 1942, box 571, exp. 515.1/85-515.1/154, FP Ávila Camacho, AGN.

39. Próspero Castro, letter to Domínguez, 11 October 1940; Domínguez, letter to SCOP, 24 October 1940, box 26, JLCNL, AGGNL; Salinas, telegram to President Ávila Camacho, 17 October 1942; Salinas, letter to the president, 22 March 1941; Waldo Romo Castro, letter to SCOP,; Domínguez, letter to SEOCCENL, 31 March 1942; Manuel Lopez (SEOCCENL), letter to Domínguez, 29 March 1942, JLCNL, box 34, JLCNL, AGGNL.

40. Próspero Castro, letter to Domínguez, 11 October 1940; Domínguez, letter to SCOP, 24 October 1940, box 26, JLCNL, AGGNL; Salinas, telegram to President Ávila Camacho, 17 October 1942; Salinas, letter to the president, 22 March 1941; Waldo Romo Castro, SCOP, 18 December 1942; Domínguez, letter to SEOCCENL, 31 March 1942; Manuel Lopez (SEOCCENL), letter to Domínguez, 29 March 1942.

41. Castro, letter to SCOP, FP Ávila Camacho, box 571, AGN; Junta Local de Caminos Nuevo León, Report to the Fourth Pan-American Congress on Highways, 11 January 1943, box 36, JLCNL, AGGNL.

42. Dirección General de Caminos del Estado (Veracruz), "Reglamento para la construcción de caminos con impuestos especiales," January 1940, box 737, AGEV.

43. Dirección General de Caminos del Estado (Veracruz), "Reglamento," box 737; Cerdán, letter to Manuel Ávila Camacho, 1942, box 958, file: 317/2 Carreteras, Coatzacoalcos, payroll information, Comité de la Carretera Coatzacoalcos–Los Tuxtlas, 1941, box 317, AGEV; Secretaría de Comunicaciones y Transporte, *Historia de las Juntas Locales de Caminos*, 295.

44. Folder: 7-3.552, Fomento, 1925–30, box 206, Secretaría de Gobierno; Donato Miranda, letter to Jorge Cerdán, 16 January 1942; Mirando, letter to Cerdán, 23 Janu-

ary 1942; Chicontepec Committee Formation Document, 8 March 1942; Rendón, letter to Miranda, 22 April 1942, box 958, AGEV.

45. Folder: 7-3.552, Fomento, 1925–30, box 206, Secretaría de Gobierno; Donato Miranda, letter to Jorge Cerdán, 16 January 1942; Mirando, letter to Cerdán, 23 January 1942; Chicontepec Committee Formation Document, 8 March 1942; Rendón, letter to Miranda, 22 April 1942.

46. Letter from Comité de Mejoras Materiales in Jesus Carranza, Veracruz to Governor Cerdán, 3 July 1941; Pablo Herrera, letter to Cerdán, 2 July 1941; letter from DCOP to the CMC Jesús Carranza, 23 July 1941; Victoriano Andrade, letter to Cerdán, 19 September 1941, box 834 and 835, AGEV.

47. Rubén Sanchez, letter to Cerdán, 28 July 1943, box 1060; Fortunato López Ríos, report on Road Conditions & Needs in Coxquihui, Veracruz, 24 November 1944, box 959; Rendón, letter to José Ramos, 13 April 1942; Aniceto González, letter to Rendón, 22 May 1942; Guillermo F. Samaniego, letter to Rendón, 20 July 1942; Rendón, letter to González, 23 July 1942; Ramos, letter to Rendón, 4 July 1942; Rendón, letter to Ramos, 6 August 1942; Rendón, letter to Ramos, 20 August 1942, box 958, AGEV.

48. Secretaría de Comunicaciones y Transporte, *Historia de las Juntas Locales de Caminos*, 301.

49. Francisco Zayas, letter to Carlos Bazán, 9 December 1941, box 958, AGEV.

50. Domínguez, letter to Grant Advertising, S.A., 2 July 1942 (multiple), JLCNL, box 33, AGGNL.

51. Domínguez, letter to Elíseo B Sánchez, 26 June 1942, JLCNL, box 33, AGGNL.

52. Domínguez, letter to Elíseo B Sánchez, 26 June 1942; Armando Arteaga y Santoyo, letter to Pablo Domínguez, 17 August 1942, JLCNL, box 33; Armando Arteaga y Santoyo, letter to Pablo Domínguez, 17 August 1942, JLCNL, box 33, AGGNL.

53. Ernestina Díaz de Leal, formal complaint to District Judge Pedro González, 30 June 1941; González, letter to the JLCNL about the Díaz suit, 30 June 1941; Pablo Domínguez, letter to González (arguing on jurisdictional grounds that the suit should target the Junta de Mejoras de Santa Catarina and not his agency), 3 July 1941, JLCNL, boxes 27–30, AGGNL, Monterrey.

54. Petra and Maria Rodríguez, letter to Pablo Domínguez, 21 June 1944, JLCNL, box 40; Armando Hoyo, letter to MP Cadereyta, Nuevo León, 25 November 1940; Arturo B. Garza y Garza, letter to Eulalio Saldívar, 11 February 1941, box 1, CLC, AGGNL.

55. Rendón, letter to Dario Soto Peredo, 23 September 1941; Soto, letter to Rendón, 2 October 1941, box 834; Rodolfo Zamora, letter to the Governor of Veracruz, 14 August 1941; Jorge Cerdán, letter to Secretary of the Junta Central de Caminos, 18 August 1941, box 834, AGEV.

56. Comité Pro-Abastecimiento de Agua, letter to the Secretario de Comunicaciones y Obras Públicas, 13 August 1942, box 33, JLCNL, SCOP, AGGNL.

57. "Modificaciones al sistema de transito en la ciudad," *El Porvenir*; José Peña Ávila, letter to the JLCNL, 20 June 1945, JLCNL, box 40, AGGNL. For a detailed study of policy responses to transit issues in northern Mexico, see Bess, "'Neither Motorists nor Pedestrians Obey the Rules.'"

58. José Peña Avila, letter to the JLCNL, 20 June 1945, JLCNL, box 40; letter from residents of Paso de la Loma to Domínguez, 9 July 1945, JLCNL, box 40, AGGNL.

59. "Importante Recepción al Lic. Arturo B. de la Garza a su llegada a los Ramones, N.L.," *El Porvenir*, 16 April 1943.

60. See the chapter "La segunda Guerra Mundial y la industrialización acelerada," in Cárdenas Sanchez, *El largo curso de la economía mexicana*, 493–567.

61. Conflictos Obreros, Monterrey, Nuevo León (Camioneros), exp. 702/4579 and exp. 521.7/29, FP Manuel Ávila Camacho, AGN. These files contain significant correspondence with the government by union groups that supported the strike and business representatives who opposed it.

62. Conflictos Obreros, Monterrey, Nuevo León (Camioneros), exp. 702/4579 and exp. 521.7/29.

63. "Es incuestionable que estas huelgas son un caso típico de penetración y predominio comunista, pues las tácticas de luchas puestas en acción, así lo revelan; además, de que en todas aquellas ocasiones en que se ha estados a punto de llegar a un acuerdo, la intervención de Campa, De León, Castillo y otros de los Representantes de filiación comunista, han destruido esas posibilidades."

64. Conflictos Obreros, Monterrey, Nuevo León (Camioneros), exp. 702/4579 and exp. 521.7/29.

65. Conflictos Obreros, Monterrey, Nuevo León (Camioneros), exp. 702/4579 and exp. 521.7/29.

66. Conflictos Obreros, Monterrey, Nuevo León (Camioneros), exp. 702/4579 and exp. 521.7/29.

67. Conflictos Obreros, Monterrey, Nuevo León (Camioneros), exp. 702/4579 and exp. 521.7/29; Cárdenas Sánchez, *El largo curso de la economía mexicana*.

68. Ávila Camacho, VI Informe de Gobierno, 1 September 1946, IP, DC, DSI (2006), 386, 393.

5. "Those Who Do Not Look Forward Are Left Behind"

The quote in the chapter title is from an editorial published in *El Norte* to describe support for modernization policies in Mexico once World War II ended, "Después de la guerra: Quien no mira adelante, atrás se queda," 3 May 1942. Portions of this chapter appeared in the *Journal of Transport History*.

1. Domínguez's husband, Arturo Garza, was no relation to the governor. "Un eslabón más en la industrialización efectiva de México," *El Porvenir*, 16 February 1949.

2. "Mejorarse la industria de autotransportes en México," *El Porvenir*, 16 February 1949.

3. "Después de la guerra," *El Norte*.

4. "Después de la guerra," *El Norte*, emphasis added.

5. "Después de la guerra," *El Norte*.

6. "EE.UU-México: Cooperación Económica," *El Tiempo*, 23 July 1943.

7. "Transcendental discurso del presidente Gral. Manuel Ávila Camacho en el banquete al Presidente F. Roosevelt" and "Histórico discurso del Presidente Franklin D. Roosevelt," *El Porvenir*, 21 April 1943.

8. Pro-road committee Cerro Azul, letter to Miguel Alemán, 27 January 1947, Archivo Clasificado, AGEV, Xalapa; Secretaría de Comunicaciones y Transporte, *Historia de las Juntas Locales de Caminos*, 301.

9. Pro-road committee Cerro Azul, letter to Miguel Alemán, 27 January 1947; Secretaría de Comunicaciones y Transporte, *Historia de las Juntas Locales de Caminos*, 301.

10. Secretaría de Obras Públicas, *Caminos y Desarrollo*, 59.

11. Martínez, *El despegue constructiva de la Revolución*, 44.

12. Martínez, *El despegue constructiva de la Revolución*, 44.

13. Garza Lorenzo, *amparo*, 3 August 1950, Tercera Sala; CNCV, *amparo*, 27 August 1951; Junta Central de Conciliación y Arbitraje del Estado de Chihuahua, *amparo*, 3 April 1951, Seminario Judicial, AHSCJN.

14. Transportes Monclova-Villa Frontera, *amparo*, 14 June 1954, Segunda Sala, Seminario Judicial, AHSCJN.

15. Secretaría de Obras Públicas, *Caminos y Desarrollo*, 59, 74.

16. "Mexico Hopes U.S. Will Lend Capital," *New York Times*, 26 February 1947; "Texts of Addresses of Alemán and Truman in Mexico," *New York Times*, 4 March 1947.

17. "México tiene un crédito abierto para carreteras," *El Norte*, 16 April 1949.

18. "Comprensión y ayuda entre México y EE.UU," *El Porvenir*, 26 February 1949.

19. Virginia Lee Warren, "Mexico Expedites Highway Projects," *New York Times*, 3 February 1947; Miguel Alemán, I Informe de Gobierno, 1 September 1947, 32–34, and II Informe de Gobierno, 1 September 1948, IP, CD, DSI (2006), 63–65; Kim, "Destiny of the West."

20. Jose Osorio Cruz, Chicontepec municipal president, letter to the governor, 17 August 1948, and Alfonso Ruiz Galindo, letter to Governor Carvajal, 10 December 1948, exp. 317/82, box 1561, AGEV; Melquiades Sánchez, letter to Secretario General de Gobierno, 24 December 1948, exp. D-1/021(x-50)/7428, box 1, CNCV; Pedro Reyna, letter to Governor Morones Prieto, 13 November 1948, exp. D1/021(x-50)7428, box 1, CNCV, AGGNL.

21. Armando Castillo Franco, *Contestación al Tercer Informe de Gobierno*, 1 September 1949, IP, DSI, Mexico City, 143.

22. Roland Goodman, "Better Motoring in Mexico: Road-Building Program Speeds New Routes for Summer Use," *New York Times*, 5 June 1949.

23. Roland Goodman, "New Developments in Mexico," *New York Times*, 12 October 1952.

24. Notaria N. 48 (Automotriz O'Farrill, S.A.), 25 February 1942, vol. 212, no. 14142, F. 210, and Notaria N. 48 (Automotriz O'Farrill, S.A.) 27 February 1942, vol. 210, no. 14160, F.190, Acervo Histórico del Archivo General de Notarias; González de Bustamente, *Muy Buenas Noches*, xxiii, 1–4; Paxman and Fernández, *El Tigre*, Kindle edition.

25. For some pro-road articles and advertisements in *Novedades*, see "El plan de recuperación nacional tiene pleno desarrollo en Zacatecas" (touting the importance of road building), 22 September 1948; "Arterias del comercio" (road-building advertisement from national organization for Chambers of Commerce in Mexico), 19 August 1949; "Se evitarán los desperfectos en las carreteras," 4 August 1950; "Kilómetros de 750 mts." (motor tourism advertisement), 3 August 1950; "Viaje a Guadalajara" (bus tourism advertisement), 4 August 1950; "Tecolutla" (motor tourism advertisement), 8 August 1950; "Cien mil turistas Americanos son esperados en la meta de la etapa final de la Carrera Panamericana," 16 August 1951; "Participación de la IRF en Congreso de Carreteras," 15 August 1952.

26. "Importancia nacional de la construcción de caminos," *Novedades*, 11 August 1950.

27. Comité Pro-Carretera Monterrey-García, letter to Miguel Alemán, 17 December 1946, exp. 515/.1/31, box 187, FP Miguel Alemán, AGN, Mexico City.

28. Comité Pro-Carretera Monterrey-García, letter to Miguel Alemán, 17 December 1946.

29. Roberto Amorós, letter to Jenaro Garza Sepúlveda, 27 December 1946, exp. 515/31, FP Miguel Alemán, AGN.

30. Arturo B. de la Garza, "Memorándum al presidente de la Republica," 9 December 1946, exp. 515.1/9, FP Miguel Alemán, AGN.

31. Asociación Nacional de Clubes de Leones de la Republica Mexicana, letter to Miguel Alemán, 7 January 1947, FP Miguel Alemán, AGN, for information on the assassination attempt against Ávila Camacho, see Quintana, *Maximino Ávila Camacho and the One-Party State*.

32. De la Garza, letter to Alemán outlining complaint with Tamaulipas, 7 December 1946, exp. 6615/48; De la Garza, letter to Alemán, 2 December 1948, exp. 565.4/884, FP Miguel Alemán, AGN.

33. "Proyecto de plan sexenal de caminos para el Estado de Nuevo León, 1949–1955," box 1, CLC, AGGNL.

34. "Construcción de carreteras," exp. 3783 S/F, Monterrey, box 1, CNCV, AGGNL.

35. Eleuterio Salinas, letter to Secretario de Gobierno, Nuevo León, 21 December 1949, exp. D-1/021 (x-24)/19602, CNCV, AGGNL

36. E. Salinas, letter to SEGOB, 21 December 1949, exp. D-1/021 (x-24)/19602, CNCV; Cámara Nacional de Comercio, letter to Alemán (about highway construction to Piedras Negras), 29 July 1949, exp. 515.1/19, FP Miguel Alemán, AGN; Margarito Raymundo Salinas, letter to Secretario General de Gobierno, Nuevo León, 5 December 1949, exp. D-1/021 (x-24)/19602, box 1, CNCV, AGGNL; "Banco Mercantil del Norte, S.A.," in *BNamericas: Business Insight in Latin America*, http://www.bnamericas.com/company-profile/en/banco-mercantil-del-norte-sa-institucion-de-banca-multiple-banorte (accessed 5 June 2016).

37. Melquiades Sánchez, letter to Secretario General de Gobierno, 24 December 1948, exp. D-1/021(x-50)/7428, box 1, CNCV, AGGNL.

38. Pedro Reyna, letter to Governor Morones Prieto, 13 November 1948, exp. D1/021(x-50)7428, box 1, CNCV, AGGNL.

39. Pedro Reyna, letter to Morones Prieto, 19 November 1949, box 1, Comisión Nacional de Caminos Vecinales, AGGNL.

40. Pedro Reyna, letter to Morones Prieto, 19 November 1949.

41. Roberto A. Cortés, letter to Pablo Domínguez, 7 December 1949, Comisión Nacional de Caminos Vecinales 1921–59, box 1, AGGNL; "Villa de Santiago llamada a ser importante centro de turismo," *El Norte*, 11 May 1950.

42. Comité Cerro Azul, letter to Miguel Alemán, 27 January 1947, AGEV.

43. Comité Cerro Azul, letter to Miguel Alemán, 27 January 1947.

44. Secretaría de Comunicaciones y Transporte, *Historia de las Juntas Locales de Caminos*, 301.

45. Jose Osorio Cruz, Chicontepec municipal president, letter to the governor, 17 August 1948; Alfonso Ruiz Galindo, letter to Governor Carvajal, 10 December 1948, exp. 317/82, box 1561, AGEV.

46. Alemán, I Informe de Gobierno, 1 September 1947, 33.

47. Eduardo Díaz, letter to Gustavo Rocha, 22 March 1948, exp. 317/52; SCOP, Departamento de Investigaciones y Laboratorios, "Informe de Análisis de Suelos," 22 March 1948, exp. 473.21, box 1561, AGEV.

48. Eduardo Díaz, letter to Ruiz Cortines, 31 May 1948, exp. 112.1/8.8, box 1561; Ángel Carvajal, Gustavo Rocha, and Carlos Hernández y Hernández, contract for road construction, 1 July 1948, exp. 317/32 [68], box 1559, AGEV.

49. Eduardo Díaz, letter to Ruiz Cortines, 31 May 1948; Ángel Carvajal, Gustavo Rocha, and Carlos Hernández y Hernández, contract for road construction, 1 July 1948.

50. Luis Rendón, Luis Rendón Jr., and Rafael Téllez Muñoz, letter to DCOP, 10 October 1946, one in a series for a road-building bid, exp. 317/0, box 1377, AGEV.

51. Carvajal, Rocha, and Hernández y Hernández, contract for road construction, 1 July 1948.

52. SCOP, Departamento de Investigaciones y Laboratorios, "Informe de Análisis de Suelos," 22 March 1948; SEGOB, Sistema de Información Municipal, Enciclopedia de los Municipios, http://www.inafed.gob.mx/work/enciclopedia/EMM30veracruz/index .html (accessed 14 January 2017).

53. Carlos Hernández y Hernández, personnel payroll (lista de raya), Conejos-Huatusco Highway, 16–20 July 1948, exp. 317/32, box 1559, AGEV.

54. Gustavo Rocha, letter to Paulino Ceballos, 4 September 1948, exp. 317/32, box 1559, AGEV.

55. DCOP, personnel payroll, Pánuco-Tempoal road, 1–15 May 1948, exp. 317/24, box 1568; Hernández y Hernández, personnel payroll, Conejos-Huatusco Highway, 16–20 July 1948; Carvajal, Rocha, and Hernández y Hernández, contract for road construction, 1 July 1948, exp. 317/32 [68], box 1559. For DCOP payroll, see file: 317/1 [39], Carretera, Laguna-Mahuixtlán, 1947, box 7631. For JCCV payroll, see file: 317/88 [82], Carretera Jalapa–San Bruno, 1947, box 7632, AGEV.

56. "Labor callada, pero fecunda es la del Lic. Carvajal," El Dictamen, 20 September 1949.

57. "Emotivo homenaje de despedida al Lic. Ángel Carvajal en Huatusco," El Dictamen, 13 November 1950.

58. Castillo Franco, Contestación al Tercer Informe de Gobierno, 1 September 1949, 143; "Mejorarse la industria de autotransportes en México," El Porvenir.

59. "Importancia nacional de caminos," Novedades, August 1948; "El plan de recuperación nacional," Novedades, September 1948; "Cien mil turistas Americanos son esperados," Novedades, August 1951; "Participación . . . en Congreso de Carreteras," Novedades, August 1952.

60. "Los contratistas de caminos se afilian a la Asoc. Mexicana," Novedades, 20 August 1949.

61. "No ha sido suspendido el tránsito en la carretera de Los Tuxtlas," El Dictamen, 3 October 1950; "Urgente necesidad de arreglar una carretera en San Andrés Tuxtla," El Dictamen, 10 October 1950; "Serios daños causó el ciclón en amplia zona del Estado" El Dictamen, 12 October 1950; "En pésimas condiciones se halla la carretera a San Andrés Tuxtla," El Dictamen, 3 November 1950; Grady Norton, "Hurricanes of the 1950 Season," Monthly Weather Review (January 1951), Weather Bureau Office, NOAA, Miami, FL, 12.

62. "Está destrozada la carretera a Paso del Toro," *El Dictamen*, 11 November 1950; "Reconstrucción del camino a Los Tuxtlas," *El Dictamen*, 12 November 1950; "Intensa labor de reconstrucción de caminos," *El Dictamen*, 15 November 1950.

63. Nacional Financiera, *50 años de revolución mexicana en cifras*, 100; Secretaría de Obras Públicas, *Caminos y Desarrollo*, 69–71.

64. William P. Carney, "10,000 in Mexico Protest Price Rise," *New York Times*, 22 August 1948; Sydney Gruson, "Mexicans Regain Pride in Regime," *New York Times*, 31 December 1953; Krauze, *La presidencia imperial*, 191–95.

6. Charting the Contours

1. Fondo Comisión Monetaria, box 894; Fondo Dirección General de Investigaciones Políticas y Sociales, Serie Informes de Extranjeros Expedientes Provisionales, box 739, AGN, Mexico City.

2. In *Made in Mexico* Susan Gauss discusses how Mexican leaders applied industrialization policy to create spaces of accommodation for political rivals, including the commercial and conservative elites of Monterrey.

3. Fulwider, "Driving the Nation," 17–19.

4. Rubin, "Decentering the Regime," 86.

5. Daniel Newcomer has argued that revolutionary authorities in León successfully made political inroads with conservative opponents through urban reconstruction programs framed as economic modernization ("The Symbolic Battleground," 63).

6. Ángel Lascuraín y Oso, "La verdad abre paso," *El Norte*, 17 August 1952.

7. Adolfo López Mateos, "Second Report to the Government," IP, 70, and "Fourth Report to the Government," IP, 228; Gómez Ibáñnez and Meyer, *Going Private*, 99.

8. Luis Echeverría, "Third Report to the Government," 1 September 1973, IP, 198.

9. Carlos Salinas, "Sixth Report to the Government," 1 November 1994, IP, 399.

10. Felipe Calderón, "Un país más y mejor conectado," 30 November 2012, http://calderon.presidencia.gob.mx/2012/11/un-pais-mas-y-mejor-conectado/ (accessed 9 June 2015).

11. Bartra, *Guerrero Bronco*, 117–38; Reyes Morales et al., "Características de la migración internacional en Oaxaca y sus impactos en el desarrollo regional," 197–207; Hogenboom, *Mexico and the NAFTA Environment Debate*, 20.

12. Daniel Hernandez, "Calderon's War on Drug Cartels: A Legacy of Blood and Tragedy," *Los Angeles Times*, 1 December 2012, http://articles.latimes.com/2012/dec/01/world/la-fg-wn-mexico-calderon-cartels-20121130 (accessed 7 June 2016); David Alire Garcia and Todd Eastham, "Mexican Gangs Block Roads, Torch Vehicles in Western Mexico," *Chicago Tribune*, 25 August 2012, http://articles.chicagotribune.com/2012-08-25/news/sns-rt-us-mexico-crimebre87p010-20120825_1_torch-vehicles-sinaloa-cartel-mexico-city (accessed 17 January 2017).

Appendix A

1. In historical terms, it can be difficult to adjust for inflation, as no single consumer price index (CPI) has been published for the period of time covered in this study. Not until 1969 did the Bank of Mexico begin publishing its annual CPI. INEGI has recorded data in the *anuarios estadísticos* (annual statistics reports), but they are not uniform as

different base years are used to make calculations. To create a unified CPI, I used two data sets for the *Índice de precios de mayoreo de la Ciudad de México* (Index for Wholesale Prices in Mexico City), which cover the periods 1928–41 and 1936–59, with 1929 and 1935 as the respective base years. Using 1935 as the common denominator between the two indices, I applied the rule of three (a/b*100) to create an adjusted index that combined the separate INEGI data with 1929 as its base year. With this adjusted index (see table 1), I calculated for inflation. I have also included the original indices published by INEGI for those interested in reviewing my calculations (tables 2 and 3).

BIBLIOGRAPHY

Archives

Archivo General de la Nación (AGN), Mexico City.

Archivo General del Estado de Veracruz (AGEV), Xalapa.

Archivo General del Gobierno de Nuevo León (AGGNL), MonterreyArchivo Histórico de PEMEX (AHP), Mexico City.

Archivo Histórico de la Suprema Corte de la Justicia de la Nación (AHSCJN), Mexico City.

Biblioteca Capilla Alfonsina, Universidad Autónoma de Nuevo León, Hemeroteca, Monterrey.

Biblioteca Miguel Lerdo de Tejada, Hemeroteca, Mexico City.

Fideicomiso Archivos Plutarco Elías Calles y Fernando Torreblanca, Mexico City.

Hemeroteca Nacional de la Universidad Autónoma de México, Mexico City.

Published Works

Aboites, Luis, and Engracia Loyo. "La construcción del nuevo Estado, 1920–1945." In *Nueva historia general de México*, 595–652. Mexico City: Colegio de México, 2010.

Ballent, Anahi. "Ingeniería y Estado: La red nacional de caminos y las obras públicas en la Argentina, 1930–1943." *Historia, Ciências, Saúde-Manguinhos*, 15, no. 3 (2008): 827–47.

Bartra, Armando. *Guerrero Bronco: Campesinos, Ciudadanos, Guerrilleros en la Costa Grande*. Mexico City: Ediciones, 2000.

Bachelor, Steven J. "Toiling for the 'New Invaders': Autoworkers, Transnational Corporations, and Working-Class Culture in Mexico City, 1955–1968." In *Fragments of a Golden Age: The Politics of Culture in Mexico Since 1940*, edited by Gilbert M. Joseph, Anne Rubenstein, and Eric Zolov, 273–326. Durham NC: Duke University Press, 2001.

Berger, Dina, and Andrew Grant Wood, eds. *Holiday in Mexico: Critical Reflections on Tourism and Tourist Encounters*. Durham NC: Duke University Press, 2010.

Berger, Dina, and Andrew Grant Wood. "Introduction: Tourism Studies and the Tourism Dilemma." In *Holiday in Mexico: Critical Reflections on Tourism and Tourist Encounters*, edited by Dina Berger and Andrew Grant Wood, 1–20. Durham NC: Duke University Press, 2010.

Bess, Michael K. "'Neither Motorists nor Pedestrians Obey the Rules': Transit Law, Public Safety, and the Policing of Northern Mexico's Roads, 1920s–1950s." *Journal of Transport History* 37, no. 2 (2016): 155–74.

———. "Revolutionary Paths: Motor Roads, Economic Development, and National Sovereignty in 1920s and 1930s Mexico." *Mexican Studies/Estudios Mexicanos* 32, no. 1 (2016): 56–82.

———. "Routes of Conflict: Building Roads and Shaping the Nation in Mexico, 1941–1952," *Journal of Transport History* 35, no. 1 (2014): 78–96.

Booth, Rodrigo. "Turismo, Panamericanismo e Ingeniería Civil: La Construcción del Camino Escénico entre Viña del Mar y Concón, 1917–1931." *Historia* 47, no. 2 (2014): 277–312.

Cárdenas Sanchez, Enrique. *El largo curso de la economía mexicana: De 1780 a nuestros días.* Mexico City: Fondo de Cultura Económica, 2015.

Coatsworth, John. *Growth against Development: The Economic Impact of Railroads in Porfirian Mexico.* DeKalb IL: Northern Illinois Press, 1981.

———. "Railroads, Landholding, and Agrarian Protest." *Hispanic American Historical Review* 54, no. 1 (1974): 48–71.

Drowne, Kathleen Morgan, and Patrick Huber. *The 1920s.* Westport CT: Greenwood Press, 2004.

Fowler Salamini, Heather. *Agrarian Radicalism in Veracruz, 1920–1938.* Lincoln: University of Nebraska Press, 1978.

Freed, Libbie. "Network of (Colonial Power): Roads in French Central Africa after World War I." *History and Technology* 26, no. 3 (2010): 203–23.

Freeman, J. Brian. "Driving Pan-Americanism: The Imagination of a Gulf of Mexico Highway." *Journal of Latino–Latin American Studies* 3, no. 4 (2009): 56–68.

———. "'La carrera de la muerte': Death, Driving, and Rituals of Modernization in 1950s Mexico." *Studies in Latin American Popular Culture* 29 (2011): 2–23.

———. "'Los hijos de Ford': Mexico in the Automobile Age, 1900–1930." In *Technology and Culture in Twentieth-Century Mexico*, edited by Araceli Tinajero and Freeman, 214–32. Tuscaloosa: University of Alabama Press, 2013.

Fulwider, Benjamin J. "Driving the Nation: Road Transportation and the Post-Revolutionary State, 1925–1960." PhD dissertation, Georgetown University, 2009.

Gauss, Susan. *Made in Mexico: Regions, Nation, and the State in the Rise of Mexican Industrialism, 1920s–1940s.* University Park: Penn State University Press, 2011.

Gómez Estrada, José Alfredo. *Gobierno y casinos: El origen de la riqueza de Abelardo L. Rodríguez.* Mexico City: Universidad Autónoma de Baja California and Instituto Mora, 2007.

Gómez Ibáññez, José A., and John Robert Meyer. *Going Private: The International Experience with Transport Privatization.* Washington DC: Brookings Institution Press, 1993.

González de Bustamente, Celeste. *Muy Buenas Noches: Mexico, Television, and the Cold War.* Lincoln: University of Nebraska Press, 2012.

Haen, Nora. *Fields of Power, Forests of Discontent: Culture, Conservation, and the State in Mexico*. Tucson: University of Arizona Press, 2005.

Hart, John M. *Empire and Revolution: The Americans in Mexico since the Civil War*. Berkeley: University of California Press, 2006.

Hogenboom, Barbara. *Mexico and the NAFTA Environment Debate: The Transnational Politics of Economic Integration*. Utrecht, The Netherlands: International Books, 1998.

Joseph, Gilbert M., and Daniel Nugent, eds. *Everyday Forms of State Formation: Revolution and the Negotiation of Rule in Modern Mexico*. Durham NC: Duke University Press, 1994.

Joseph, Gilbert M., and Daniel Nugent. "Popular Culture and State Formation in Revolutionary Mexico." In *Everyday Forms of State Formation: Revolution and the Negotiation of Rule in Modern Mexico*, edited by Gilbert M. Joseph and Daniel Nugent, 3–23. Durham NC: Duke University Press, 1994.

Kim, Jessica. "Destiny of the West: The International Pacific Highway and Pacific Borderlands, 1929–1957." *Western Historical Quarterly* 46, no. 3 (2015): 311–33.

Knight, Alan. "The Characters and Consequences of the Great Depression in Mexico." In *The Great Depression in Latin America*, edited by Paulo Drinot and Alan Knight, 213–45. Durham NC: Duke University Press, 2014.

———. "Introduction: *Caciquismo* in Twentieth-Century Mexico." In *Caciquismo in Twentieth-Century Mexico*, edited by Alan Knight and Will Pansters, 150. London: Institute for the Study of the Americas, 2005.

———. "Weapons and Arches in the Mexican Revolutionary Landscape." In *Everyday Forms of State Formation: Revolution and the Negotiation of Rule in Modern Mexico*, edited by Gilbert M. Joseph and Daniel Nugent, 24–68. Durham NC: Duke University Press, 1994.

Knight, Alan, and Will Pansters, eds. *Caciquismo in Twentieth-Century Mexico*. London: Institute for the Study of the Americas, 2005.

Krauze, Enrique. *La presidencia imperial: De Manuel Ávila Camacho a Carlos Salinas de Gortari*. Mexico City: Tusquets Editores, 1997.

Kuntz-Ficker, Sandra. *Empresa extranjera y mercado interno: El Ferrocarril Central Mexicano, 1880–1907*. Mexico City: El Colegio de México, 1995.

Martínez, María Antonia. *El despegue constructivo de la Revolución: Sociedad y política en el alemanismo*. Mexico City: Centro de Investigaciones y Estudios Superiores en Antropología, 2004.

Matthews, Michael. *The Civilizing Machine: A Cultural History of Mexican Railroads, 1876–1910*. Lincoln: University of Nebraska Press, 2014.

Meyer, Jean. "Revolution and Reconstruction in the 1920s." In *Mexico since Independence*, edited by Leslie Bethell, 201–40. Cambridge, UK: Cambridge University Press, 1991.

Moore, Rachel. *Forty Miles from the Sea: Xalapa, the Public Sphere, and the Atlantic World in Nineteenth-Century Mexico*. Tucson: University of Arizona Press, 2011.

Mora-Torres, Juan. *The Making of the Mexican Border: The State, Capitalism, and Society in Nuevo León, 1848–1910*. Austin: University of Texas Press, 2001.

National Council for Science and Environment. *Encyclopedia of Earth*. Edited by Cleveland Cutler. Washington DC: Environmental Information Coalition, 2009.

Nacional Financiera. *50 años de revolución mexicana en cifras*. Mexico City: Nacional Financiera, 1963.

Nava Negrete, Alfonso. "Derecho de las Obras Públicas en México." Biblioteca Jurídica Virtual del Instituto de Investigaciones Jurídicas. Mexico City: Universidad Nacional Autónoma de México: http://biblio.juridicas.unam.mx/libros/6/2688/16.pdf. Originally published in *Actualidad y perspectivas del derecho público a fines del siglo XX*, vol. 2. Madrid: Editorial Complutense, 1992.

Newcomer, Daniel. "The Symbolic Battleground: The Culture of Modernization in 1940s León, Guanajuato." *Mexican Studies/Estudios Mexicanos* 18, no. 1 (2002): 61–100.

Niblo, Stephen. *Mexico in the 1940s: Modernity, Politics, and Corruption*. Wilmington DE: SR Books, 2000.

———. *War, Diplomacy, and Development: The United States and Mexico, 1938–1945*. Wilmington DE: SR Books, 1995.

O'Flaherty, Coleman A. *Highways: The Location, Design, Construction and Maintenance of Pavements*. 4th edition. London: Butterworth-Heinemann, 2002.

Paxman, Andrew, and Claudia Fernández. *El Tigre: Emilio Azcárraga y su imperio Televisa*. 2nd edition. Mexico City: Grijalbo, 2013.

Paz, María Emilia. *Strategy, Security, and Spies: Mexico and the U.S. as Allies in World War II*. University Park: Pennsylvania State University Press, 1997.

Pussetto, Cintia Smith, Nancy Janett García Vázquez, and Jesús David Pérez Esparza. "Análisis de la ideología empresarial regiomontana." *CONfines* 4, no. 7 (2008): 11–25.

Quintana, Alejandro. *Maximino Ávila Camacho and the One-Party State: The Taming of Caudillismo and Caciquismo in Post-Revolutionary Mexico*. New York: Rowman and Littlefield, 2010.

Reyes Morales, Rafael G., Alicia Sylvia Gijón Cruz, Antonio Yúnez Naude, and Raúl Hinojosa Ojeda. "Características de la migración internacional en Oaxaca y sus impactos en el desarrollo regional." In *Nuevas Tendencias y desafíos de la migración internacional México-Estados Unidos*, edited by Raúl Delgado Wise and Margarita Favel, 195–221. Mexico City: Universidad Autonoma de Zacatecas, 2004.

Rubin, Jeffrey. "Decentering the Regime: Culture and Regional Politics in Mexico." *Latin American Research Review* 31, no. 3 (1996): 85–126.

Santiago, Myrna I. *The Ecology of Oil: Environment, Labor, and the Mexican Revolution, 1900–1938*. New York: Cambridge University Press, 2006.

Saragoza, Alex M. *The Monterrey Elite and the Mexican State, 1880–1940*. Austin: University of Texas Press, 1988.

Scott, James C. *Seeing Like a State: How Certain Schemes to Improve the Human Condition Have Failed*. New Haven CT: Yale University Press, 1998.

Secretaría de Comunicaciones y Transporte. *Historia de las Juntas Locales de Caminos, 1933–1980*. 1st edition. Mexico City: n.d.

Secretaría de Obras Públicas. *Caminos y Desarrollo: México 1925–1975*. Mexico City: Unidad Editorial, 1975.

Van Hoy, Teresa. *A Social History of Railroads in Mexico: Peons, Prisoners, and Priests*. New York: Rowman and Littlefield, 2008.

Vaughan, Mary Kay. *Cultural Politics in Revolution: Teachers, Peasants, and Schools in Mexico, 1930–1940*. Tucson: University of Arizona Press, 1997.

Waters, Wendy. "Remapping Identities: Road Construction and Nation Building in Post-Revolutionary Mexico." In *The Eagle and the Virgin: Nation and Cultural Revolution in Mexico, 1920–1940*, edited by Mary Kay Vaughan and Stephen Lewis, 221–42. Durham NC: Duke University Press, 2006.

———. "Remapping the Nation: Road Building as State Formation in Post-Revolutionary Mexico, 1925–1940." PhD dissertation, University of Arizona, 1999.

Weber, Eugen. *Peasants into Frenchmen: The Modernization of Rural France, 1870–1914*. Stanford: Stanford University Press, 1976.

Wolfe, Joel. *Autos and Progress: The Brazilian Search for Modernity*. Oxford: Oxford University Press, 2010.

INDEX

In The Mexican Experience series

Seen and Heard in Mexico: Children and Revolutionary Cultural Nationalism
Elena Jackson Albarrán

Railroad Radicals in Cold War Mexico: Gender, Class, and Memory
Robert F. Alegre
Foreword by Elena Poniatowska

Mexicans in Revolution, 1910–1946: An Introduction
William H. Beezley and Colin M. MacLachlan

Routes of Compromise: Building Roads and Shaping the Nation in Mexico, 1917–1952
Michael K. Bess

Radio in Revolution: Wireless Technology and State Power in Mexico, 1897–1938
J. Justin Castro

San Miguel de Allende: Mexicans, Foreigners, and the Making of a World Heritage Site
Lisa Pinley Covert

Celebrating Insurrection: The Commemoration and Representation of the Nineteenth-Century Mexican Pronunciamiento
Edited and with an introduction by Will Fowler

Forceful Negotiations: The Origins of the Pronunciamiento *in Nineteenth-Century Mexico*
Edited and with an introduction by Will Fowler

Independent Mexico: The Pronunciamiento *in the Age of Santa Anna, 1821–1858*
Will Fowler

Malcontents, Rebels, and Pronunciados: *The Politics of Insurrection in Nineteenth-Century Mexico*
Edited and with an introduction by Will Fowler

Working Women, Entrepreneurs, and the Mexican Revolution: The Coffee Culture of Córdoba, Veracruz
Heather Fowler-Salamini

The Heart in the Glass Jar: Love Letters, Bodies, and the Law in Mexico
William E. French

"Muy buenas noches": Mexico, Television, and the Cold War
Celeste González de Bustamante
Foreword by Richard Cole

CPSIA information can be obtained
at www.ICGtesting.com
Printed in the USA
LVOW11s1419081117
555504LV00001B/36/P